Lecture Notes
in Control and Information Sciences   212

Editor: M. Thoma

W0051282

Springer-Verlag London Ltd.

Arturo Sanchez

Formal Specification and Synthesis of Procedural Controllers for Process Systems

Springer

Series Advisory Board

Author

Arturo Sanchez, Dr
Centre for Systems Process Engineering, Imperial College, London SW7 2BY, UK

ISBN 978-3-540-76021-4

British Library Cataloguing in Publication Data
Sanchez, Arturo
  Formal specification and synthesis of procedural
  controllers for process systems. - (Lecture notes in
  control and information science)
  1.Process control
  I. Title
  629.8
  ISBN 978-3-540-76021-4   ISBN 978-3-540-44430-5 (eBook)
  DOI 10.1007/978-3-540-44430-5
Library of Congress Cataloging-in-Publication Data
A catalog record for this book is available from the Library of Congress

Typesetting: Camera ready by author

69/3830-543210 Printed on acid-free paper

To Andres and Paulina

Contents

List of Figures

List of Tables

Preface

During the last twenty years, process (i.e. chemical, petrochemical, pharmaceutical and food) industries have experienced a dramatic transformation on production paradigms. This has been the response to domestic and international demands for more efficient and cleaner processes and wider variety of higher quality, lower cost products. As a result more flexible production environments have emerged, characterised by frequent changes in the product recipes, production modes and equipment configuration. Thus, the role of event–driven activities in process plants has grown in importance. Examples are sequence operations, handling of alarms, emergency procedures and equipment interlocking. In automated plants, these activities are normally driven by control systems which are able to handle logical aspects of the operation. A variety of names is used to identify these devices depending on the engineering discipline being applied. For instance, terms such as logic, batch or switching controller can be found in the automotive, process or telecommunications industries. Here, the term *procedural controller* will be used to broadly identify this type of device and its application will be focussed on the process industries. Synthesis and analysis of procedural controllers is a key stage in process automation because these controllers ensure both a correct normal operation of the process and a safe handling of abnormal conditions accounting for expected safety and environmental hazards. Although procedural controllers are heavily used in practice (e.g. implemented in programmable logic systems), limited theoretical frameworks exist to support their formal design and analysis. The need for such a framework as well as formal methods and tools for synthesis, verification and validation has been stressed by academics and industrialists (Schuler *et al.*, 1991).

Dedicated control architectures have also been proposed to accommodate the increasing number of organisation and coordination activities for specific types of process systems (Akamatsu *et al.*, 1989; Cott and Macchietto, 1989). This has unveiled the urgent need to use formal techniques and tools to design, install and maintain these systems (Wilkins, 1992). In practice, the difficulties during design and implementation[1] and even regrettable failures of computer-based control systems, not only in the process industry[2] (McIver, 1994), have confirmed this need. Moreover, it is accepted by the industrial community that mathematical techniques may be exploited, not only to enhance the operation of such systems, but also to improve the quality, efficiency and development effort of control software at a lower cost (Benveniste and Astrom, 1993). Batch processing is an area in which these considerations are particularly important due to its discrete operation nature. Recent surveys of batch process operation and engineering have been presented by Crooks (1992) and Sawyer (1993).

Methods and techniques for software specification and verification originally from computing sciences have been used in the study of control issues in event–driven systems with limited results (e.g. Ostroff, 1989b; Benveniste *et al.* 1989; Brooks *et al.* 1989; Ravn *et al.* 1993). The main obstacles have been the conceptual model of system on which these methods and techniques are based and the role of the controller in a closed-loop system. Most approaches fail to consider the controlled process as a system with its own particular dynamics in which external intervention is done only through external commands. For instance, a control system issuing a command to open a valve is not a guarantee that the valve will actually open. Nevertheless, the potential applicability of specific tools in process systems has been undoubtedly demonstrated (Yamalidou *et al.*, 1990; Caines and Wang, 1991; Moon *et al.*, 1992).

An important theoretical leap on the use of theory initiated in computing sciences is the Supervisory Control Theory proposed by Wonham *et al.* (Ramadge and Wonham, 1987b; Ramadge and Wonham, 1987a; Wonham and Ramadge, 1987). In this theory, the process and its desired behaviour are modelled as finite state machines (*FSMs*) in which transitions are labelled as con-

---

[1] In August, 1993, British Nuclear Fuels pinpointed 2,400 faults in an early version of the software for its £1.8 billion nuclear fuel reprocessing plant at Sellafield (Marks, 1992).

[2] The London Ambulance Service came to an almost stand still for 36 hours during the last week of October in 1992, after a failure in the "overcomplicated" dispatch control system recently acquired (The Guardian, 1992; Arthur, 1992).

trollable or uncontrollable. Using these models a device, termed a supervisor, may be synthesised. A supervisor acts upon the process disabling controllable transitions in such a way that it guarantees the process is restricted to a controllable set of trajectories satisfying the desired behaviour. The Supervisory Control Theory gives some general foundations for the study of control issues for this class of systems. This same generality however, has so far precluded its practical use mainly due to two aspects:

- The difficulty in constructing realistic models of the process and the desired behaviour specifications to be imposed upon the process.

- The limited range of controllable systems covered by the theory and of theoretical concepts to aid in the controllable behaviour synthesis.

Moreover, Supervisory Control Theory was not conceived to consider the enforcing of control actions upon a process. A comprehensive control framework for process systems must consider this aspect. Control is exerted on a process by enforcing actions and not just disabling their occurrence.

This work capitalises on Supervisory Control Theory and software specification techniques and investigates the synthesis of procedural controllers for process systems that can be modelled as state–transition structures. Limitations posed by the current frameworks are highlighted and solutions are presented having in mind the needs of process systems. Emphasis is given to controllers amenable to formal proofs of correctness. The study is restricted to systems described by *FSMs* in which only qualitative time aspects are included. In recent years these systems have been identified as a class of Discrete Event Systems (DESs). The problem posed is that of finding a feedback controller that carries out a given operating procedure or enforces a given behaviour upon the system. In other words, the controller must be able to use feedback information from the process to execute control commands on the process as a function of the system responses.

Outline

In order to establish a control framework for event–driven systems based on Supervisory Control Theory, the three aspects mentioned above (modelling limitations, extension of controllable systems and the enforcement of control actions upon the process) must be addressed and resolved.

Chapter 1 briefly presents the origins of procedural control, the relation between this type of control and DESs and how DES tools have been used in the past in process control. Then, it reviews previous work in the control and specification of event–driven systems in which logic aspects are of particular relevance. Several control paradigms are presented from which the Supervisory Control Theory is identified as the most suitable basis for a synthesis method. Regarding specification modelling, the prescriptive power and solid algebraic foundations of logic formalisms make them an ideal candidate for modelling behaviour specifications. The translation of natural language statements into formal and compact representations is facilitated, emphasising the occurrence of states or transitions, or their combination, as required.

The limitations mentioned previously for constructing realistic models are treated in the following two chapters. In chapter 2, the state–transition structure to be used as the modelling tool in the *FSM* domain is presented. The goal is to develop process and specification models with a unified and formal treatment. Structural characteristics are defined using algebraic properties to handle a maximum of information of particular states or sequences of states while maintaining compact representations. It is shown how process models are constructed in an incremental and modular fashion using standard operators from Automata theory adapted to account for the structural characteristics of the proposed modelling representation. Two simple examples are used to illustrate the construction of models from their elementary components. Having set up a general framework for modelling, attention is directed towards the capture and model construction of behaviour specifications. Within Supervisory Control Theory, a very efficient way of dealing with specifications related to logic invariant properties, such as reaching buffer capacities or avoiding forbidden states, has been introduced (Ramadge and Wonham, 1987a; Kumar *et al.*, 1991b). This method is used in this work to deal with forbidden states of operation. However, no methods have been proposed for the construction of specification models of realistic size involving dynamic behaviour, such as the sequencing of events. It has always been assumed that these specification models are given in advance. Moreover, the construction of such models becomes extremely difficult and very prone to errors even for simple specifications. In chapter 3, a formalism to perform this task based on Linear Temporal Logic (LTL) is explored. Dynamic specifications are modelled as LTL formulas and then translated into *FSMs*. The LTL formulas are constructed using Predicate Logic connectives and three basic LTL operators, next ($\bigcirc$), always ($\square$) and eventually ($\Diamond$). The LTL formulas obtained are characterised by predefined syntactic structures called *logic schemes*. The translation process then becomes the construction of *FSMs* as logic models in which the LTL formulas are true. For the efficient algorithmic translation, homomorphisms are given which make use of lattice–based structures which allow the characterisation and compact representation of large sets of states. Examples are presented for the construction of typical specifications and their translation into the *FSM* domain.

Once in possession of a consistent modelling framework and tools to construct process and specification models, control issues of event–driven systems are considered. In order to lay down theoretical foundations required for further development, chapter 4 gives a detailed account of the controllability and supervision aspects of Supervisory Control Theory. Firstly, the feedback supervisory structure proposed in the Supervisory Control Theory is presented and discussed. This supervisor can only disable controllable transitions that the process can execute. It is not capable of enforcing actions upon the process. Subsequently, the definitions of controllability and proper supervisor are presented, together with conditions of existence for a supervisor and methods for determining the largest set of controllable closed–loop process behaviour existing under a given supervisor. The Supervisory Control Theory defines a system as controllable if no uncontrollable transition forces the system to leave the set of controllable behaviour. This enforces conservative conditions on the supervisor synthesis by assuming that any uncontrollable transition in a given system state will be executed. Consequently such behaviour must be captured in advance by the corresponding behaviour specifications, making their construction very difficult in the early stages of supervisor design. This is resolved by relaxing the concept of controllability to permit the occurrence of controllable transitions under circumstances in which uncontrollable transitions can occur as well. The above allows to identify such transitions and thereupon, it becomes feasible to either prescribe the behaviour generated from these transitions within the specification model or consider them as potential causes of uncontrollable behaviour. A supervisor satisfying this relaxed controllability is defined, and a calculation method is presented for finding the largest set of controllable closed–loop process behaviour associated with this supervisor.

At this stage, the necessary theoretical tools have been established with which to attack the third problem mentioned above, that is the need for mechanisms to enforce the execution of control actions. As a first proposal, chapter 5 introduces the notion of a controller which is able to decide upon control actions to be executed in the process using a feedback arrangement. This controller is obtained by restricting the behaviour enabled by the supervisor in such a way that at most one controllable transition exists in each controller state. A procedure is developed for the synthesis of such a controller and two examples illustrate its use. The first example deals only with the construction of a controller for normal process operation. The model size is small enough to present pictorially each step of the synthesis procedure. The second example expands the results of the first. The objective is to construct a controller not only for normal operation, but also for conditions identified in the first example as possible sources of abnormal operation.

The controllers synthesised in the previous chapter are unable to resolve the execution of control commands when explicit time–related aspects are involved (e.g. how long must the controller wait for a process response before issuing a given control command). These limitations are explored and resolved in chapter 6 using two more examples for which dynamic models are constructed considering continuous and discrete aspects. The models are implemented in a general purpose simulator together with the synthesised controllers. The ability of the controllers to drive the system is shown by simulation. The chapter closes with a summary of the material presented.

Finally, an epilogue presents the material covered in the book, closing with a discussion on the applicability of the proposed methodology.

Acknowledgements

I wish to express my gratitude to Prof. Sandro Macchietto for his advice and support along this work. Dr. Guillermo Rotstein was most helpful for clarifying many of the ideas presented here. Dr. Costas Pantelides suggested the g*PROMS* model for implementing the synthesised controllers. Nicholas Alsop had the patience to read the manuscript at different stages of development.

Chapter 1

The Role of Procedural Control

1.1 Introduction

Procedural control is associated with both continuous and event–driven features of a process. The continuous part of a process is usually composed of time–driven events that can be modelled by continuous variables using differential–difference and/or algebraic equations. In addition, the discrete part may be described by symbolic, logical or numerical values, the changes in which are usually event–driven. Unfortunately, dynamics not described by continuous variables (or their time discretisation) are poorly understood. This is reflected in the current lack of solid theoretical frameworks for their study. In recent years the discrete/continuous nature of dynamic processes has been recognised as a fundamental aspect due to the proliferation of discrete–operation systems and the introduction of computer–based process components. This gives rise to complex dynamics and new control issues in event–driven systems, which are receiving considerable attention in the control community.

In the early 80's, the name Discrete Event Systems (DESs) was coined to identify the wide variety of systems in which event–driven dynamics are of main relevance. Disciplines to study these event–driven systems and their control aspects have emerged, borrowing at first techniques and tools mainly from Computing Sciences and Operations Research. At present, some areas such as Parametric Analysis of Stochastic Processes (Ho and Cao, 1992; Rubenstein and Shapiro, 1992) and Supervisory Control Theory (Wonham, 1989) are well established. Good introductory discussions on DESs can be found in Varaiya and Kurzhanski (1988), the IEEE special issue in DESs, vol. 77, no. 1, 1989, Ramadge and Wonham (1989) and, specifically in Chemical Process Control, Yamalidou *et al.* (1990). This chapter looks at the current state–of–the–art in the area. Section 1.3 surveys work in DESs in Chemical Process Control, while section 1.4 considers in more detail control issues related with DESs. Existing control paradigms are reviewed focussing on Supervisory Control Theory, which has attracted considerable attention in the last ten years due to its solid theoretical foundations. The use of Temporal Logic (TL) as an analysis and specification tool is surveyed in section 1.5. Temporal Logic (TL) is a suitable candidate for the construction of desired behaviour models due to its assertive power and sound algebraic foundations. Finally, the chapter closes with a summary and some conclusions.

1.2 Procedural Control in Process Systems

The term procedural control was introduced in the ISA standards for batch control (ISA-dS88.01, 1994) as the device that "directs equipment-oriented actions to take place in an ordered sequence in in order to carry out a process-oriented task". Procedure is understood as "the strategy for carrying out a processing action". Having in mind the above, the term procedural control will be associated here to the execution of procedural actions at any level in the control hierarchy.

Procedural control originally emerged in the manufacturing and telecommunication industries (Michel, 1990). The first application was reported in 1969 as a substitute for massive relay grids needed for sequential operation in the automotive industry. In chemical process industries, the use of procedural control became popular during the 70's with the introduction of electronic programmable devices known as Programmable Controllers (PCs)[1]. Nowadays PCs are more than simple sequencers and include many other non procedural control functions that are making them as popular in chemical processes as distributed control systems (Schuler *et al.*, 1991).

Procedural control has been historically outside the mainstream of the control discipline and little attention had been given to it by the academic community. The lack of a theory that can be used by practitioners has been highlighted

---

[1] Programmable Logic Controller, which is a term frequently used in the chemical process industries, was coined by Allen–Bradley for its first programmable controller developed in 1969.

by the industrial community which has relied on control and instrumentation companies for problem solving. Many times, solutions are found heuristically or intuitively, especially in the batch processing area (Schuler *et al.*, 1991). In other areas such as digital systems and switching theory, specific techniques and tools to design the controller "logic" have been developed (Harrison, 1965; Potton, 1973; Comer, 1984). These techniques and tools are mainly based on Boolean Algebra and Automata Theory. On the other hand, in recent years more general specification tools for software control systems such as STATE-CHARTS (Harel, 1987) and GRAFCET (IEC, 1988) have gained broad acceptance in manufacturing industries. Also, tools for software verification have been used for the design of Very Large Scale Integrated circuits (VLSI) (Dill, 1989) and sequential control of chemical processes (Moon *et al.*, 1992; Hiranaka and Nishitani, 1994; Moon and Macchietto, 1994). However, the results obtained from the use of these tools have stressed even more the need for a solid theory for controller synthesis and analysis which, to a great extent, is still missing.

1.3   DESs in Chemical Process Control

After thirty years of intense interest in continuous processes, the need of innovative productions paradigms and the emergence of suitable frameworks for their study (modelling formalisms, programming paradigms and computational techniques) have attracted the attention of the Chemical Engineering academic and industrial community towards DESs. Areas of early incursion of DESs theory and formalisms include discrete event process modelling (Preisig, 1989; Yamalidou *et al.*, 1990; Hanisch, 1992,1993) and control (Yamalidou and Kantor, 1991), fault analysis and diagnosis (Prock, 1991; Yamalidou *et al.*, 1990) and rescheduling (Yamalidou *et al.*, 1992). In the following paragraphs, works related with the modelling and analysis of DESs in chemical process control are examined. The section closes with a survey of controller synthesis for DESs in chemical processes. Table 1.1 presents a summary of the main references.

1.3.1   Modelling and Analysis.

Traditionally, the analysis of the dynamic performance and validation of DESs in Chemical Engineering is done with problem–specific tools and focusses on modelling issues using simulation as the main technique. Several good general–purpose packages exist, such as GPSS (Schriber, 1974), SLAM II (Pritsker, 1986), SIMULA (Birtwistle, 1979) and the discrete–continuous chemical pro-

Reference	Modelling	Analysis	Controller Synthesis
Preisig (1989)	Automata		Intuitive controller building
Yamalidou et al. (1990)	Temporal Logic		Control policies. Logic reasoning
Yamalidou et al. (1990)	Calculus of Events	Cause–event analysis in HazOp studies	
Yamalidou et al. (1990)	Minimax Algebra	Timing analysis in batch plants	
Yamalidou et al. (1990)	Petri Nets	Analysis by simulation of batch operations	
Yamalidou and Kantor (1991)	High Level Petri Nets		On–line control. MILP
Hanisch (1992)	Petri Nets		
Moon et al. (1992)	FSM/CTL	Verification of sequential controllers	
Hiranaka and Nishitani (1994)	FSM/CTL	Verification of batch operations	
Moon and Macchietto (1994)	FSM/CTL	Verification of batch operations	

Table 1.1: Summary of DESs references in Chemical Process Control.

cess package g*PROMS* presently under development (Barton and Pantelides, 1994). In the last few years, the use of DES modelling tools has permitted to explore the use of different analysis methods amenable to a more formal approach. Yamalidou *et al.* (1990) presented a survey of four methods including Temporal Logic, Calculus of Events, Minimax Algebra and Petri Nets, and demonstrated their possible application to different areas of chemical process control. Temporal Logic was used to synthesise a control strategy for a simple process. This will be discussed in detail in the next subsection. Event Calculus was applied in the analysis of cause–event relations for hazard evaluation techniques. *HazOp* techniques were used to construct a database model composed of relationships between *HazOp* guide words and changes in process equipment and parameters. The database was implemented in PROLOG. Minimax algebra models were applied to the study of timing in a multipurpose batch plant. Finally, Petri Nets were employed to model and analyse the activities in a batch facility. Time was assigned to places modelling activities. Transitions between activities were assumed to be instantaneous. Control places were introduced to resolve conflicts by shared resources. Firings were controlled by an enabling vector which determines the earliest time that an enabled transition can fire. Yamalidou and Kantor (1991) introduced a systematic approach for the use of high level Petri Nets to model pipe/valve networks (Rivas and Rudd, 1974). Using a modular approach, rules were developed to model tasks, equipments and shared resources as well as their connections. Coloured tokens described different compounds and valves. Inhibitor arcs were introduced to avoid (token) accumulation in places. Although the proposed rules lead to modelling restrictions (e.g., valves do not admit two–directional flow), their use facilitates the model construction task.

The use of Petri Nets for the modelling of coordination activities has been proposed elsewhere (Wang and Saridis, 1990). In chemical process control, Hanisch (1992) explored the modelling of such coordination activities in batch systems using three types of nets: condition/event, place/transition and predicate/transition. Examples are provided to illustrate their use. Models were constructed manually and no exogenous information was allowed in the net. It is also claimed that Petri Nets is a suitable tool for the design and performance analysis of supervisory controllers because the net possesses sufficient information to perform control tasks but this topic is not elaborated further.

Recently, TL and verification methods were used for the analysis of sequential controllers and alarm systems by Moon *et al.* (1992). Moon and Macchietto (1994) applied them for checking logical inconsistencies in coordination

activities of batch processes. In these two works, the system under analysis is
modelled as a labelled graph. Assertions to be checked in the system regarding
the dynamic behaviour are written using a branching Time TL, named Computational Tree Logic* (CTL*), proposed in Emerson and Halpern (1986). A
model checker (Clarke *et al.*, 1986) is employed to check the logical consistency
between the set of assertions and the system model. If an inconsistency is
found, the model checker shows where the model fails to satisfy the assertion
in question. In the present implementation of the model checker, the checking
process is stopped after the first inconsistency is found. Once the analysis with
the model checker is finished, changes must be introduced in the labelled graph
so as to redress any violations highlighted by the model checker. At present,
no formal methodology has been proposed to perform such modifications. A
more systematic modelling approach was presented by Hiranaka and Nishitani
(1994) who describe process and controller models using labelled graphs. Each
of the elements of a given process is clearly identified, but the actual model
construction task is still left to the user. In order to simplify the modelling of
complex systems, a way of dealing with incomplete information in model states
was introduced. It is assumed that a given state in a graph can take values of a
given domain as well as a symbol in which all the values are considered simultaneously. This is similar to the *covering symbol* introduced here in chapter 2.
However, the definition of this symbol and its use were introduced by Hiranaka
and Nishitani (1994) in a rather informal way. Although this description of
the modelling components may be of some use, the actual rationale behind the
construction of the graph is still far from being systematic. In summary, in
order to take advantage of these techniques in realistic problems, methods to
construct models and to introduce changes in process and controller models
after verification are yet to be proposed.

1.3.2   Synthesis

A very limited number of works has been published concerning controller synthesis for DESs in chemical processes. The use of *FSMs* in chemical processes
for the construction of procedural controllers was first documented by Preisig
(1989). A Mealy machine was produced in a modular fashion using submachines representing different tasks to be executed in the process. The tasks
included normal process operation, alarm/fail/emergency procedures and the
interaction among them. An intuitive analysis of the submachines was performed to obtain the minimal representation realising the same behaviour. The
procedure was not further systematised and no implementations were reported.

Yamalidou *et al.* (1990) used a Linear Temporal Logic framework for the synthesis of control strategies for a simple process. Temporal Logic formulas were used to model the process behaviour, responses to control commands and operational specifications to be enforced on the process. Propositional Logic statements were given representing initial and final system conditions that must be satisfied. Using the process and operational specification models, a Temporal Logic formula was deduced representing the control strategy that fulfils the initial and final conditions. The example was simple enough to allow the reader to corroborate the logic followed in the generation of the TL formula. No systematic procedure was presented to perform the reasoning.

The only published work in the chemical process literature related to the synthesis of controllers is by Yamalidou and Kantor (1991). "On–line" control policies were determined for pipe/valve networks modelled as high level Petri Nets. Modelling aspects were discussed in subsection 1.3.1. The control policy consisted of the firing sequence in the Petri Net. It was obtained as the result of a MILP problem minimising the number (or the cost) of control actions required to reach a final marking given an initial marking. The Petri Net was used to generate some of the constraints. These included the *enabling condition*, which determines which transition can be fired at a given instant, and the *inhibitor constraints* generated by the inhibitor arcs introduced in the Petri Net to avoid token accumulation in some places. Other constraints associated with the operation, which could be represented as Boolean expressions in conjunctive normal form, were also included. These were identified as *operational constraints* and were given by elements of the Petri Net marking vector describing valves and components. The Boolean expressions were then translated into linear inequalities. The modelling of the process as well as the calculation of the "optimal control policy" is done in a systematic fashion. However, from a safety point of view, the use of controllers that can be fully verified and validated in advance is preferable.

1.4   Control of Discrete Event Systems

When dealing with a particular DES, it is common to use different modelling tools according to the task to be performed. Highly assertive formalisms are utilised for process specification or prescription while descriptive formalisms are preferred when algorithmic issues are of main interest. For instance, Temporal Logic (TL) is frequently employed as a specification language while frameworks such as Finite State Machines (*FSMs*) or Petri Nets are used to describe the process model. This is a well established practice in software engineering from which the process engineering community has adopted incipient techniques.

Controller design is a modelling–analysis–synthesis exercise. Most of the control paradigms use, to certain extent, both types of formalisms relying heavily on either logic or *FSM*–based formalisms. In the following pages, the control paradigms surveyed are classified according to which modelling framework plays the main role either for the analysis of the closed–loop system or the synthesis of the actual control device. First, logic–based controllers are presented, followed by those frameworks using Automata or Languages as main modelling formalisms, and particularly, focussing on Supervisory Control Theory. Finally, some works using dual–language approaches are surveyed. A references summary is presented in table 1.2.

1.4.1   Logic–based Frameworks

The use of LTL formulas for the analysis and synthesis of procedural controllers was first explored by Fusaoka *et al.* (1983). The process dynamics and behaviour specifications of a simple process (a boiler) were modelled using Manna and Pnueli's LTL formulas (Manna and Pnueli, 1983) and then translated into a particular type of graph, called ω–graph, in which the reasoning task was performed. Behavioural properties such as "eventuality" (occurrence of certain events in the future), "stability" (reaching a certain state after a finite time) and "observability" (the values of the system are known by the controller) were expressed as LTL formulas. A generic control structure was defined in advance (in this case, an IF/THEN structure). The particular controller satisfying an individual behaviour specification was constructed assuming that the intersection of the set of behaviours of the system and controller was given by the behaviour generated by the ω–graph of the specification model. Then, the controller was reconstructed from this intersection. No framework was given to formalise the results and apparently the topic was not explored further. The suitability of the Manna and Pnueli's LTL framework for systems with a physical meaning was not discussed. This was done by Thistle and Wonham (1986) who proposed a more rigorous use and adapted its proof system for the analysis of discrete–event controllers and closed–loop dynamics. The differences from Manna and Pnueli's proof system lay mainly in the introduction of "next event" symbols to treat the immediate occurrence of transitions and in the omission of generalisation rules allowing the use of LTL formulas to describe trajectories, even when such formulas are not always true (for instance, when modelling a process with physical meaning, for a given formula f, it is not

Reference	Process	Specifications	Subject
Fusaoka et al. (1983)	LTL/graphs	LTL/graphs	controller synthesis
Thistle and Wonham (1986)	LTL	LTL	closed-loop analysis
Caines and Wang (1991)	FSMs/TL	LTL	inference (feedback) control
Ramadge and Wonham (1987b)	FSMs	FSMs	state feedback supervision
Ramadge and Wonham (1987a)	FSMs	Predicate Logic/FSMs	state feedback supervision
Cieslak et al. (1988)	FSMs	FSMs	supervision under partial observations
Lin and Wonham (1990)	FSMs	FSMs	supervision under partial observations
Zhong and Wonham (1990)	FSMs	Predicate Logic/FSMs	decentralised and hierarchical supervision
Kumar et al. (1991a)	FSMs	FSMs	model-based state supervision
Kumar et al. (1991b)	FSMs	Predicate Logic/FSMs	model-based state supervision
Balemi et al. (1993)	FSMs	FSMs	state-supervision of process systems
Brave and Heymann (1993)	STATECHARTS	FSMs	hierarchical supervision
Chung et al. (1992)	FSMs	FSMs	limited lookahead supervision
Ostroff (1989b)	Extended FSMs	LTL	real-time control modelling and analysis
Ostroff (1989a)	Extended FSMs	LTL	real-time controller synthesis
Passino and Antsaklis (1990)	Petri Nets	CTL	closed-loop system behaviour
Knight and Passino (1990)	FSMs	LTL	closed-loop system behaviour

Table 1.2: Summary of references in DESs control.

always true that $\Box f$). Two examples were presented in which the open–loop plant dynamics, closed–loop specifications and controller specifications were described using LTL formulas. Then, using the proof system, the verification of the controller specifications as part of the closed–loop specifications and plant dynamics was performed. According to the authors, the translation of natural language specifications into LTL formulas is more straightforward and the LTL formulas obtained are easier to understand as high level specifications. They also recognised that the use of Language Theory for modelling specifications can be difficult and error–prone.

Logic systems and automatic theorem proving as a control paradigm have been used by Caines and Wang (1991) for on–line control. The logic controller presented, named COCOLOG (COnditional COntrol LOGic) comprises two parts. The first part contains process dynamics information and desired closed loop behaviour as a set of TL formulas which can be subject to completeness proofs (i.e. any other formula not contained in the original set of formulas prescribing system behaviour can be derived from the original set). The process dynamics are modelled first as a *FSM* which is then translated into first order logic formulas. The second part consists of an inference machine equipped with suitable inference rules. An initial observed state and a control goal specifying the desired performance written as a TL formula and target state must also be given. Then, the inference machine (an automatic theorem prover) finds the sequence of control actions that must be imposed upon the system to satisfy the control goal. In other words, it finds how the final state can be deduced from the initial state using the provided set of logic formulas. The set of TL formulas is updated with new information about the dynamics of the system every time an observation is made. In this way, the set of formulas improves its knowledge of the system dynamics. No formal framework to characterise the closed–loop behaviour or the controller is given. Regarding the actual implementation of the controller, it is assumed that the required inference results are available instantaneously at each clock instant. Presently, this represents a serious drawback for any on–line implementation because the current technology of theorem provers do not support such fast calculations. An implementation of COCOLOG is currently being explored in a specifically built theorem prover (Caines *et al.*, 1993).

1.4.2   Supervisory Control Theory

Perhaps the DES control formalism that has received most attention in the last ten years is Supervisory Control Theory, introduced by P. J. Ramadge and W.

M. Wonham in three seminal papers: Ramadge and Wonham, 1987a, 1987b; Wonham and Ramadge, 1987. In this theory, a feedback mechanism, termed *supervisor*, disables controllable actions in the process under supervision in order to satisfy given behaviour specifications. This fundamental framework covers the following topics:

- A general formalism to model a process and its supervisor using *FSMs*. The system behaviour is characterised as a language, understood as the set of all sequential occurrences of transitions, while a *FSM* is designed as the generator of such a language by the sequential execution of the *FSM* transitions.

- Characterisation of control–related concepts such as "control action" and "controllability" as well as conservative conditions for the existence of supervisors to achieve a given controllable behaviour.

- Methods for the calculation of the controllable behaviour of a given system.

- The use of either Automata and Language Theories or Predicate Calculus to focus the approach from either an event–based or state–based point of view respectively.

A good overview of this framework can be found in Wonham (1988) and a considerable list of references in Ramadge and Wonham (1989). The basis of Supervisory Control Theory can be divided into three sections:

- Controllability and supervision.

- Supervision under partial observations.

- Hierarchical and decentralised supervision.

Each of these three sections is surveyed in the following paragraphs. This work concerns only the first, with a more detailed account presented in chapter 4.

1.4.2.1   Controllability and Supervision

1.4.2.1.1   Basic Definitions and Modelling Issues.   The first part of the theory defines the basic nomenclature and deals with modelling issues. It introduces the concept of controllability as well as conditions for the existence of supervisors (Ramadge and Wonham, 1987b; Ramadge and Wonham, 1987a; Wonham and Ramadge, 1987).

Models capture only sequential behaviour and are constructed using a class of *FSMs* describing two types of behaviour:

- The system behaviour, which represents all the possible trajectories that the set of transitions can generate.

- The marked behaviour, which comprises those trajectories that reach a set of states of specific interest (e.g., completion of tasks, initial or final states, alarm points) identified as marked states.

Transitions are defined as controllable if they represent events that can be disabled by external intervention. Otherwise, the transitions are uncontrollable.

The construction of a supervisor using appropriate operators from Automata and Language theories requires that each *FSM* involved considers the complete set of transitions in order to obtain the required behaviour. To achieve this, self–loops attached to each state were informally introduced (Wonham, 1988). A self–loop contains all the transitions that are not relevant to the *FSM* behaviour in the state to which the self–loop is attached. Balemi *et al.* (1993) formalised the use of self–loops by introducing operators to obtain specific language projections.

In the original Supervisory Control Theory, the process model is understood as a *FSM* generator of controllable or uncontrollable transitions. The supervisor is modelled as another *FSM* together with a state feedback function that maps the supervisor state into control patterns. Such control patterns disable the occurrence of only controllable transitions (Ramadge and Wonham, 1987b). The notion of supervisor has migrated from the first ideas of feedback function and control patterns to simpler and more realistic representations. For instance, supervision by synchronous composition (Heymann, 1990) of process and supervisor was introduced by Kumar *et al.* (1991a). Here the process and supervisor are assumed to run in full synchronisation under the set of transitions. Therefore, if some of the transitions do not occur in the supervisor, then they also cannot occur in the process. Following these ideas, the design of a model–based supervisor was discussed such that the closed-loop language is equal to the language accepted by the supervisor. In other words, the language generated by the interaction of process and supervisor is given by the intersection of process and supervisor languages, $L(P) \cap L(S)$. Assuming this language as the supervisor language, it is easy to see that its intersection with the process language gives as a results the supervisor itself: $L(P) \cap (L(P) \cap L(S)) = L(P) \cap L(S)$.

1.4.2.1.2   Controllability and Existence of Supervisors. The introduction of the notion of controllability and conditions of existence for supervisors that guarantee a particular, although conservative, behaviour on the
system was one of the most important achievements of the basic Supervisory
Control Theory. The most relevant definitions of supervisors for this work are
those of *complete* and *proper* supervisors. A *complete* supervisor with respect
to a given process is one that guarantees that none of the uncontrollable transitions, whose precedent transitions are part of the controllable behaviour, are
prevented from occurring in the closed–loop system. Ramadge and Wonham
(1987b) developed necessary and sufficient conditions for its existence. Simpler proofs were presented by Kumar *et al.* (1991a) assuming the process and
supervisor execute synchronously the same set of transitions. To ensure that
the controller does not block the process from reaching desired marked states,
the concept of a *proper* supervisor was introduced in Ramadge and Wonham
(1987b). Sufficient conditions for its existence were presented in Wonham and
Ramadge (1988). This will be discussed in section 2.5.

1.4.2.1.3   Calculations of Controllable Behaviour. Methods for calculation of the maximum set of controllable behaviour (supremal controllable
language) were first discussed in Wonham and Ramadge (1987) in which two
methods were presented. The first method is based on the iterative calculation
of a controllable language by a fixed–point operator. The second method is
specifically tailored for closed languages. It is based on the iterative checking
of controllability constraints on the *FSM* realising a given language. Kumar *et
al.* (1991a) proposed a non iterative algorithm for closed languages based on
an alternative definition of controllability. Apparently, this algorithm seems to
be more advantageous as it avoids iteration, although both algorithms have an
equivalent complexity of O(mn) where $m = \max|L_{Process}|$ and $n = \max|L_{Specs}|$
(Wonham and Ramadge, 1988; Kumar *et al.*, 1991a). Both calculation methods present problems of state explosion. Closed formulas for the calculations of
supremal controllable languages have also been proposed (Brandt *et al.*, 1990).
Balemi *et al.* (1993) presented an efficient implementation based on Binary Decision Diagrams (Brace *et al.*, 1990) that reduces the impact of computational
complexities in the calculations. Although this implementation does not resolve
the problems of state explosion, the size of the demonstrated examples is such
that calculations for realistic examples seems to be feasible. Originally, this
calculation technique was used by Burch *et al.* (1990) for the model checking
of structures of up to 10^{20} states.

1.4.2.1.4   State and Event Approaches.   Language representation em-
phasises the occurrence of events. To consider circumstances in which the supervision objectives are given in terms of states and to circumvent the problem
of extensive state manipulation, tools from Predicate Logic have been used for
the reformulation of the definition of controllability and conditions of existence
of controllable languages and supervisors in terms of predicate statements and
predicate transformers (Ramadge and Wonham, 1987a; Kumar *et al.*, 1991b).
Predicate statements are a powerful way of characterising state sets and predicate transformers can describe the state evolution of a *FSM*, thus facilitating
the specification of logic invariant properties, such as mutual exclusion properties, forbidden states, finite counting and buffering (Kumar *et al.*, 1991b) with
which closed–loop system must comply.

1.4.2.1.5   Modular Synthesis of Supervisors.   Ramadge and Wonham
(1987a) explored the synthesis of supervisors and its complexity issues using a
modular approach for properties that can be associated to Predicate Logic (PL)
statements (logic invariant properties). A generalised approach was presented
by Wonham and Ramadge (1988) and Wonham (1988). The objective was to
reduce the computational complexity of the synthesis task. Sufficient conditions
for the existence of modular solutions were given. This topic is elaborated
further in chapter 4.

1.4.2.2   Supervision under Partial Observations

The theory presented above can be expanded to consider supervision under
partial observations (Cieslak *et al.*, 1988; Wonham, 1988; Lin and Wonham,
1990). Calculation methods for the maximal controllable languages are discussed in Kumar *et al.* (1991a; 1993) and closed formulas are given in Brandt
et al. (1990). Reconstruction of behaviour using output signals by inversion of
the system has been studied by Ozveren and Willsky (1992).

1.4.2.3   Decentralised and Hierarchical Supervision

Hierarchical structures are exploited within DESs to study complex systems
commonly found in control systems. Zhong and Wonham (1990) show the use of
the proposed theoretical machinery within different hierarchical levels which are
related by equivalence relations. Such relations give rise to desired partitions
on the sets of low level states in such a way that only relevant information
survives in higher level states of the system structure. A recent application

of hierarchical control using the STATECHARTS formalism (Harel, 1987) was presented by Brave and Heymann (1993).

Some other interesting topics related to Supervisory Control Theory not covered here include computational complexity issues (Tsitsiklis, 1989; Rudie and Willems, 1993) the inclusion of time (Wong–Toi and Hoffmann, 1991; Gaubert, 1993; Brandin and Wonham, 1994) and concurrency aspects (Heymann, 1990; Wang, 1993).

1.4.2.4   Limited Lookahead Supervisor

Based on Supervisory Control Theory, a restricted horizon supervisor termed Limited Lookahead Supervisor (LLS) was introduced by Chung et al. (1992). This control paradigm partially circumvents the state explosion problem by unfolding the system behaviour only N steps ahead of the current state. The treatment is not restricted to regular languages and the execution of multiple transitions is permitted. The LLS performs the necessary calculation on–line to find a locally controllable system behaviour N steps ahead of the current state. Trajectories leading to uncontrollable behaviour N steps ahead are disallowed. Conditions to find globally valid LLSs are given. Such conditions are equivalent to finding the number of steps contained in the longest uncontrollable trajectory of a given system and then using that number as the span of the calculations for the LLS. No methods to calculate the maximum uncontrollable trajectory were given. This problem was also explored by Knight and Passino (1990) in a TL context and no satisfactory answer was found. At present only specifications related to states are treated (equivalent to the logical invariant specifications).

1.4.3   Dual–language Frameworks

The use of dual–language frameworks has lead to developments able to handle more complex applications. Ostroff (1989b) developed the Real Time Temporal Logic–Extended State Machine (RTTL/ESM) framework for the analysis of real–time controlled systems. In this framework, a LTL system based on the Thistle and Wonham's (1986) modified version of LTL and an extended formalism based on *FSMs* were used to model specifications and system dynamics (process and controller) respectively. *FSMs* were extended to consider continuous variables, local clocks and concurrency by communicating channels. A proof system was developed together with decision procedures on LTL frames for reasoning on formulas. The method has proved to be promising in the analysis and verification of small examples but no realistic applications have been presented so far. Ostroff (1989a) showed how to use the RTTL/ESM framework for the synthesis of controllers under conservative conditions to satisfy "safeness" properties. First, the undesired behaviour must be identified in terms of states in the process model. Then, mechanisms are given to back–trace step by step the evolution of the model behaviour. At each back–traced step, the determination of a control policy which avoids reaching the undesirable behaviour is considered. This is repeated until a suitable control law is found. The method has not been developed further. Also, TL plus *FSMs* were used by Knight and Passino (1990) to study closed–loop system behaviour. The use of algorithmic mechanisms were explored to determine whether a TL formula holds in the system model. In a similar fashion, Passino and Antsaklis (1990) investigated the use of Computation Tree Logic (CTL) and Petri Nets. Here, the model checker taken from Clarke *et al.* (1986) was used to analyse CTL formulas in a closed–loop system, following the same line as Moon *et al.* (1992), Hiranaka and Nishitani (1994) and Moon and Macchietto (1994).

1.5   Behaviour Specification and Temporal Logic

Most control frameworks assume that a model of the desired behaviour is given in advance, either for analysis of the closed–loop behaviour or the synthesis of supervisors and/or controllers. In our experience, the construction of these models when using *FSM*-based formalisms has proved to be extremely difficult even for simple specifications. Other works reported in the literature corroborate the above (Thistle and Wonham, 1986; Brooks *et al.*, 1990). In order to facilitate the construction of behaviour specifications, more appropriate formalisms must be used in the first instance.

The formal prescription of systems has received considerable attention in Software Engineering regarding the specification, implementation, verification and validation of software systems (Jones *et al.*, 1991; Lindsay, 1988). In recent years, the study of real–time systems and related control issues have also attracted the attention of the wider systems community. Formal languages based on net formalisms (i.e. Petri Nets (Diaz and Silveira, 1983; Diaz, 1987; Reisig, 1987; Willson and Krogh, 1990)) or logic formalisms (i.e. Temporal Logic (Hale, 1989)) for description, specification and analysis of systems are widely used. Design methodologies (Klittich and Seifert, 1988; Brooks *et al.*, 1990) and software environments for real–time systems design (Benveniste *et al.*, 1989; Ghezzi and Pezze, 1993) can also be found.

In particular, Temporal Logic (TL) possesses a rich enough algebraic structure to generate simple and compact model representations while maintaining a high degree of modularity (Goldblatt, 1992). This characteristic eases the synthesis of controllers when the process is complex and/or of realistic size. TL is a nonstandard logic which has been widely used for the last fifteen years as a powerful formal reasoning tool mainly in Computing Sciences for the modelling and analysis of properties related to qualitative time such as sequentiality, eventuality or necessary occurrence of events. A good introduction to the topic is given by Goldblat (1992). Particularly in Software Engineering, TL has had a substantial impact in program verification and synthesis. Since Pnueli (1984) suggested its use as a modelling tool for the verification of concurrent programs, a great amount of related works have appeared, in which different TL frameworks and applications have been presented.

The structure of time imposes the main standard for classification of TL frameworks. The two main versions are linear and branching. In linear time each instant has a unique successor. Therefore the structures over which a linear time formula is interpreted are linear sequences. Versions of linear time TL (LTL) vary according to their use (Manna and Pnueli, 1982; Wolper, 1983; Thistle and Wonham, 1986). In branching time, each time instant may have several successors. Infinite trees are the structures over which formulas of this type are interpreted. CTL (Emerson and Clarke, 1982) and CTL* (Emerson and Halpern, 1986) are examples of branching TL. Many other variations can be found in the literature (e.g. Partial order TL which considers past operators (Pinter and Wolper, 1984) or interval logics (Halpern et al., 1983)). Other characteristics to be considered are the representation of time as either a discrete or continuous quantity and the interpretations of TL formulas at intervals of time or pointwise.

Due to its prescriptive power and formal structure, TL has found application as a prototyping (Hale, 1989) and specification language in several software development tools (Lynch, 1991; Jones et al., 1991). In particular, methods based on Logic formalisms used in computer program verification have had a considerable impact in the engineering community. These can be classified into three main approaches (Wolper, 1989):

- **Axiomatic verification.** The computer program (or the process model) as well as each of the specifications to be verified is expressed as a set of temporal formulas. A proof system is used to prove that the program formulas imply the desired properties by firstly trying to prove the logical equivalence between the specifications and the program using an axiomatic system. Minimal TL formulas are then resolved by algorithmic decision procedures. Ostroff (1989a; 1989b) used a similar approach in his RTTL/ESM framework.

- **Model Checking.** The program (or the process model) is modelled as a finite state structure while the specifications are represented by TL formulas. A model checker (an algorithmic procedure) is used to check whether model and specifications are non contradictory in a logical sense. This approach has been used by Dill (1989) in VLSI circuit design, and by Moon *et al.* (1992), Hiranaka and Nishitani (1994) and Moon and Macchietto (1994) for the verification of sequential controllers in process engineering.

- **Automata Verification.** Program as well as specifications are modelled as finite state structures. The verification is done by algorithmic techniques.

The fact that a class of programs as well as their behaviour specifications can be modelled as finite state structures suggests that they can be synthesised from a model of the specification. Attempts have been reported in the literature for deterministic as well as non deterministic programs (Manna and Wolper, 1984). Wolper (1989) presents a review on program verification and synthesis methods as well as the characterisation of models for different TLs.

1.6   Summary and Conclusions

The study of dynamics and control of systems in which discrete events play a major role has gained importance in the last fifteen years mainly in disciplines outside process engineering, such as manufacturing systems and computer sciences, in which man–made dynamic systems are particularly important. DESs in process engineering are relevant due to the current availability of highly flexible process equipment performing batch operations and the increasing number of computer–controlled systems in which interactions between process and computer give rise to complex dynamics. Safety aspects become of primary importance in the computer–controlled operation of these processes stressing the need for methods and tools for the synthesis, verification and validation of such systems. As summarised in table 1.1, most published works address modelling. The use of verification techniques has been explored but no systematised (Moon *et al.*, 1992; Hiranaka and Nishitani, 1994; Moon and Macchietto, 1994). Only two works address controller synthesis. In the first, Yamalidou *et al.* (1990) used a TL framework to synthesise TL formulas modelling control policies for small examples. Although the approach seemed promising for off–line calculations, no further work has been reported. Yamalidou and Kantor (1991) employed high level Petri Nets to determine "on–line" control policies for valve/pipe networks modelled as coloured Petri Nets. Modelling rules were proposed to systematise the model construction, and although these rules lead to modelling restrictions (e.g., valves do not admit two–directional flow), their use facilitates enormously the model construction task. The control policy consisted of the firing sequence in the Petri Net obtained as the result of an MILP problem minimising the number (or the cost) of control actions required to reach a final marking given an initial marking. The use of this "optimal control" approach makes the verification and validation of closed–loop performance difficult, which from a safety point of view, is not desirable.

In summary, the synthesis of procedural controllers for chemical processes that are provably correct is a topic receiving little attention. Theories emerging in other disciplines may prove to be highly valuable, not only in the development of theoretical foundations for a rigorous and systematic approach for the synthesis of procedural controllers but also for establishing a link with other disciplines in which specification and design engineering tools (Harel, 1987; Lindsay, 1988; Benveniste *et al.*, 1989; Hale, 1989; Jones *et al.*, 1991; Ghezzi and Pezze, 1993) are now emerging.

Table 1.2 summarises references in DESs control and supervision discussed in this chapter. Among them, Supervisory Control Theory is one of the most rigorous theories available. Supervisory Control Theory deals with what has been called "supervisory aspects" of a control system. In this context, supervision is understood as the action of maintaining the closed–loop behaviour of a given system within a space in which controllability can be guaranteed by disabling controllable actions in a feedback mode. The device used to perform the disabling operations is called a *supervisor*.

In the basic Supervisory Control Theory, process and behaviour specifications are modelled using *FSMs*. The concept of "controllable behaviour" is formally defined and calculation methods are given to find such a behaviour for a given system. Conditions for the existence of different classes of supervisors and calculation methods are also available. However, the applicability of this theory to process systems is severely limited by:

- Restrictive modelling tools. Standard *FSM* mechanisms are not well suited for the manipulation of states and transitions on a large scale.

- The complexity in the construction of the specification models for the desired behaviour required for the supervisor synthesis. It is a cumbersome and error–prone task, even for very small examples. Specification models containing precise states along with (sequences of) transitions are particularly difficult to construct.

- The restricted class of behaviour considered due to the definition of controllability. It imposes conservative conditions which restrict substantially the controllable space of a given system.

Furthermore, Supervisory Control Theory does not consider the enforcing of control actions upon a process. This is a very important aspect that has not been treated in the past and which is necessary if a framework for control of DES is to be developed using a *FSM*/Language–based approach.

Techniques from software engineering for specification and verification have been found useful in attacking the problem of specification modelling on a practical basis and to allow the shaping of a more formal framework. In particular the use of TL has proved to be effective. The algebraic foundations of TL make it suitable for use in conjunction with state–transition structures such as *FSMs*. The prescriptive capabilities of TL naturally consider the following aspects:

- Easy translation of natural language statements into TL formulas.

- Compact modelling.

- Emphasis on the occurrence of either states or events as needed.

Chapter 2

Modelling Framework

2.1 Introduction

In this chapter, the modelling tools used to construct discrete–event models are developed. The main objective is to produce a representation which is both effective for modelling complex systems, and which is amenable to formal analysis. First, the definition of a basic modelling representation is introduced in section 2.2. This is a labelled finite state machine (*FSM*) termed *a–machine*, in which each *state* is characterised by a set of *state–variables* describing the process. Section 2.3 deals with the definition of the structural properties associated with *a–machines*. Ways of comparing *states* as well as *state–variables* are developed by means of algebraic relations and partial orders. These make possible the practical manipulation of large models during the specification and synthesis tasks. They also permit the characterisation of some of these models as lattices, which are well studied algebraic systems with a compact and structured representation. The partial orders defined for *states* and *state–variables* in section 2.3 are associated with lattice structures. This makes it possible to establish mappings from Predicate and Temporal Logic formalisms, which are introduced in the next chapter for the modelling of behaviour specifications, to the modelling formalism developed in this chapter. In section 2.4, basic definitions from Lattice Theory are introduced together with some illustrative examples. Section 2.5 presents a formal characterisation of the set of trajectories generated by *a–machines* using concepts from Language Theory. Elementary definitions are given together with some language properties to be used in chapter 4 when dealing with supervisory control issues. Since *a–machines* are *FSMs*, operators from Automata Theory can also be used for their manipulation. Section 2.6 introduces two of these operators, the *asynchronous product* and the *synchronous product*, which are modified in order to consider the special *a–machine* features. These operators will be applied in model construction and the controller synthesis. Finally, in section 2.7, two examples illustrate the ideas presented in this chapter. In both examples, a model is generated which describes the discrete–event behaviour of a process using the modular definition of its elementary components as a starting point.

2.2   Model Definition

The modelling structure utilised is a labelled *FSM*, named *a–machine*. This is defined as a 7–tuple

$$M = \{Q, V^{n_v}, \Sigma, \delta, \gamma, q_0, Q_m\}$$

where

Q Set of *states* : $q \in Q$.

V^{n_v} Set of *state–variables* : $\{(v_j)_q \ ; j = 1 \ldots n_v$ (number of *state–variables* defining *state* q)$\}$.

Σ Set of transitions : $\sigma \in \Sigma$.

δ State transition function defined as a partial function $\delta : \Sigma \times Q \to Q$.

γ *State–variable* transition function, defined as a partial function $\gamma : \Sigma \times V^{n_v} \to V^{n_v}$.

q_0 Initial *state*.

Q_m Set of marked *states*.

A *state–variable* $(v_j)_q$, $j = 1 \ldots n_v$ describes an elementary component of the process (e.g. the position of an on/off valve, the status of a pump). A system is assumed to be fully described by a set of n_v *state–variables* at any given moment. For each *state–variable* v_j, a finite domain is defined from which the *state–variable* takes values. Different values for each of these *state–variables* define a particular *state*, $q \in Q$, of the *a–machine*.

For instance, consider a system composed of one on/off valve as elementary component. The *state–variable* describing this component is "valve position" and its domain is {open, closed}. Therefore, a model of this system has two *states* in which the "valve position" takes values "open" or "closed". A model of two on/off valves, v_1 and v_2 could be composed of four *states* with two *state-variables* in each *state* describing all the possible combinations of the valve positions.

In addition to the domain values, two more symbols may be assigned to a *state-variable*: ∞_j, which symbolises all the possible values that *state-variable* j can take and $\infty_j^{Le_j}$, which represents all the possible values for *state-variable* j, except those in the set Le_j. As will be shown in subsections 2.3 and 2.4, these symbols help to construct the required *a-machines* in an efficient way and to write them in a compact form.

Transitions are instantaneous events leading from one source *state* to a destination *state* and changing only one *state-variable* value. Σ is the set of the system transitions and Σ^* is the set of all system trajectories composed of transitions in Σ. The partial function $\delta : \Sigma \times Q \to Q$ defines the relation between *states* and transitions, while the partial function $\gamma : \Sigma \times V^{n_v} \to V^{n_v}$ does the same for *state-variables* and transitions. Each transition must be labelled as either controllable or uncontrollable. For instance, issuing a command to close a valve is a controllable transition but the changing of a liquid level in a tank is not.

The initial *state* q_0 is a unique *state* while the set of marked *states* Q_m distinguishes *states* of special significance for the system (e.g. *states* where operation can be held or a task is completed).

An ordered set of *state-variables* $(v_j)_q$, $j = 1 \ldots n_v$ is associated with each *state* q. Therefore, a homomorphism $\beta : Q \to V^{n_v}$ is defined such that

$$\beta[\delta(\sigma, q)] = \gamma[\sigma, \beta(q)]$$

This modelling structure is introduced because the same *state-variable* values may occur more than once in different *states* of a system. For control purposes, it is important to distinguish different occurrences in order to identify behaviour as a function of the interaction of several system components, as will be illustrated in sections 3.2 and 3.3.

The construction of a model is done incrementally. Elementary models must be generated individually for each single part of the process (e.g. valves, measuring devices, interlocks, switches). Only one *state-variable* is associated with each elementary model (e.g. "valve_position", "level_status"). A transition is always associated with a change in the *state-variable* value (e.g. if the current value of *state-variable* "valve_position" is "valve open" and transition "valve is closing" occurs, in the next *state* of the model the value of *state-variable* "valve_position" will be "valve closed"). Thus, in a more complex model generated from several elementary models, the partial function $\gamma : \Sigma \times V^{n_v} \to V^{n_v}$

defines which *state–variable* is changing under the execution of a given transition.

As mentioned previously, a value for each *state–variable* (i.e. each elementary process component) must be assigned in each *a–machine state*. During the model construction, the symbols ∞_j and $\infty_j^{Le_j}$ may be assigned to a *state–variable* to handle incomplete information. Generally speaking, ∞_j is assigned to *state–variable* v_j in those *states* in which its value is not relevant, permitting v_j to take any value from its corresponding value set. $\infty_j^{Le_j}$ specifies that *state–variable* v_j can take values exclusive of the set Le_j. In the next section, a formal way is introduced to manipulate and compare *states* and *state–variables* in which these symbols appear. For the remainder of this section, some standard definitions from Automata theory (Eilenberg, 1974) are given which will be used in chapter 5. These have been rephrased for their use in Supervisory Control Theory (Ramadge and Wonhan, 1987a; 1987b).

Definition 2.1 *Reachable state subset.*

Subset of states which can be reached from the initial state.

$$Q_r : \{q \in Q \ / \ \exists \sigma \in \Sigma^*, \ \delta(\sigma, q_0) = q\} \tag{2.1}$$

Where Σ^* is the set of all possible system trajectories (string set of the language).

Definition 2.2 *Coreachable state subset.*

Subset of states from which a marked state can be reached.

$$Q_{cr} : \{q \in Q \ / \ \exists \sigma \in \Sigma^*, \ \delta(\sigma, q) \in Q_m\} \tag{2.2}$$

An *a–machine* M is *reachable* if all *states* $q \in Q$ can be reached from q_0. M is *coreachable* if all marked *states* $q \in Q_m$ can be reached from q_0.

Definition 2.3 *Trim Machine*

A machine which is reachable and coreachable is said to be trim.

Marked *states* are of particular relevance because they indicate circumstances in which the process starts or finishes, can be halted or held. In process systems it is desirable that all the marked *states* can be reached from the initial *state* and thus, that the structures being used to model processes, specifications or controllers are trim. From definitions 2.2 and 2.3 it follows that every machine has a *trim* substructure (Eilenberg, 1974; Ramadge and Wonham, 1987b) in which the *state* set, marked *states* and δ-function are given by

$$Q_{new} \;=\; Q_r \cap Q_{cr}$$
$$Q_{mnew} \;=\; Q_m \cap Q_{new}$$
$$\delta_{new} \;=\; \delta \;:\; \Sigma \times Q_{new} \to Q_{new}$$

2.3   Structural Properties

As mentioned in the past section, the symbols ∞_j and $\infty_j^{Le_j}$ permit the handling of incomplete information and facilitate the construction of the model. The symbol ∞_j is assigned to those *state–variables* of no interest at a particular *state* and therefore they may take any value from their corresponding value set. The symbol $\infty_j^{Le_j}$ is more specific on the values that a given *state–variable* can take, the set Le_j specifying those values not permitted in v_j. The appropriate transitions must be attached to the given *state* in a self–loop to permit the occurrence of designated *state–variable* values. In order to handle incomplete information regarding these *state–variables*, partial orders on the *states* as well as on the *state–variables* are introduced. These definitions allow the formal comparison of *states* and to "cover" or "refine" the *state–variable* values on each *state* as needed.

Assume the existence of *states* i and $j \in Q$ where $(v_k)_i$ and $(v_k)_j$, $k = 1 \ldots n_v$, are the *state–variables* in *states* i and j respectively, and n_v is the number of *state–variables* in each *state*, then the following partial order is introduced:

Definition 2.4 *Covering (Refinement) (Davey and Priestley, 1990).*
State–variable $(v_k)_i$ covers state–variable $(v_k)_j$ ($(v_k)_j$ refines $(v_k)_i$) if

$$(v_k)_i = \infty_k$$

State i covers state j (j refines i) if
for $k = 1 \ldots n_v$, there exists at least one state–variable such that $(v_k)_i = \infty_k$, while for the rest of the state–variables $(v_k)_i = (v_k)_j$

The above definition introduces the need for differentiating or equating *states*. This is done as follows:

Definition 2.5 *Equality.*
State–variables $(v_k)_i$ and $(v_k)_j$ are said to be equal if they share the same value (i.e. $(v_k)_i = (v_k)_j$).
States i and j are equal if $\forall k$, $(v_k)_i = (v_k)_j$ and $(v_k)_i \neq \infty_k$

In other words, a *state-variable* $(v_k)_i$ covers *state-variable* $(v_k)_j$ if the value of $(v_k)_i$ is the covering symbol. A *state i* covers a *state j* if all the *state-variables* in *state i* either cover or are equal to their corresponding *state-variables* in *state j*. If all the *state-variables* are specified (symbol ∞_k does not appear) and are the same in both *states*, then it is said that the *states* are equal. It is important to note that if some of the *state-variables* in *state i* cover their corresponding *state-variables* in *state j* while some other *state-variables* in *state j* cover their corresponding *state-variables* in *state i*, then the covering is undefined and the *states* cannot be compared. If $(v_k)_i = \infty_k$, it is said that such a *state-variable* is covered. If all *state-variables* in *state i* are covered, then the *state* is covered. In order to deal with the covering of specific *state-variable* values using the *state-variable* value subset Le_j, another partial order is introduced. Assume again the existence of *states i* and $j \in Q$ with $(v_k)_i$ and $(v_k)_j$, $k = 1 \ldots n_v$.

Definition 2.6 *Partial Covering.*
State-variable $(v_k)_i$ *partially covers state-variable* $(v_k)_j$ *if*

$$(v_k)_i = \infty_k^{Le_i} \text{ and } (v_k)_j \notin Le_i$$

This permits a more detailed comparison of some of the *a-machine states*.
If $(v_k)_i = \infty_k^{Le_i}$ it is said that this *state-variable* is partially covered. If all *state-variables* in *state i* are partially covered then *state i* is partially covered.

2.4   Lattices and a–machines

Under certain circumstances, *a-machines* can be characterised as lattices which are algebraic systems with a well defined formal structure. This allows the definition of formal equivalences among compact representations. As will be shown in chapter 3, it also eases the construction of desired behaviour models, using Temporal Logic formulas to model process behaviour, which are then translated into the *a-machine* domain using lattice–based homomorphisms. In the following paragraphs some basic definitions related to Lattice Theory are presented. For an introduction to lattices and partial orders see Donellan (1968) and Davey and Priestley (1990).

A lattice is an algebraic system $< \Lambda, \wedge, \vee >$, where Λ is a set, $a, b \in \Lambda$, $\wedge$ and $\vee$ are binary operations on Λ which are idempotent, commutative and associative.

$$a \vee a = a \wedge a = a$$
$$a \vee b = b \vee a \ ; \ a \wedge b = b \wedge a$$
$$(a \vee b) \vee c = a \vee (b \vee c) \ ; \ (a \wedge b) \wedge c = a \wedge (b \wedge c)$$

A lattice possesses a *partial order*, $\leq$, in which every pair of elements $a, b \in \Lambda$ has a greatest lower bound (glb) $a \wedge b$ and a least upper bound (lub) $a \vee b$.

If $a \leq c$ then $a \wedge c = a$ and $a \vee c = c$. Making c of the form $c = a \vee b$, a lattice satisfies the absorption laws:

$$a \wedge (a \vee b) = a; \ a \vee (a \wedge b) = a$$

A *bounded* lattice is a finite lattice where the least and the greatest elements are glb and lub. In a bounded lattice $< \Lambda, \wedge, \vee, glb, lub >$, $b \in \Lambda$ is a complement of $a \in \Lambda$ if

$$a \wedge b = glb \text{ and } a \vee b = lub$$

Note that the definition is symmetric in a and b. Therefore a is a complement of b if b is a complement of a.

A lattice is *complemented* if every element in the lattice set Λ has at least one complement in Λ.

A lattice is *distributive* if for any $a, b, c \in \Lambda$

$$a \vee (b \wedge c) = (a \vee b) \wedge (a \vee c)$$
$$a \wedge (b \vee c) = (a \wedge b) \vee (a \wedge c)$$

These two equalities are equivalent (Tremblay and Manohar, 1975). Therefore, in order to check whether a lattice is distributive it is sufficient to verify one of these two equalities for all possible combinations of the lattice elements. In a distributive lattice, if an element $a \in \Lambda$ has a complement, then it must be unique.

A *Boolean* lattice is a complemented distributive lattice. A typical example of Boolean lattices is $< \rho(S), \cap, \cup, \sim, \emptyset, S >$, where S is an nonempty set, $\rho(S)$ is the power set of S, $\cap$ and $\cup$ are the set intersection and union operations, $\sim$ is the complement of any subset $A \subseteq S$, such $\sim A = S - A$. Another example is the algebraic system $< S, \wedge, \vee, \neg, \text{F}, \text{T} >$ where S is a set of statement formulas with n variables, $\wedge, \vee, \neg$ are the conjunction, disjunction and negation operators and F and T denote the formulas that are false or true. The model of the burner system, discussed in section 2.7 and shown there in fig. 2.6, is also a Boolean lattice. All the possible system *states* are considered together with all the possible transitions between *states*. The partial order is given by the actual structure of the graph. The glb is the *state* labelled as 1 in the fig. 2.6, in which all the *state–variables* take the initial values defined by the elementary models. The lub is given by the *state* in which all the *state–variables* take the complement value of those in *state* 1, labelled as *state* 16 in the figure. It can be seen that each *state* has a unique complement, and the whole structure is distributive.

Boolean lattices and their associated partial order can be efficiently exploited to obtain compact models. In fact, definition 2.4 and particularly definition 2.6 are useful in the construction of models that contain all the possible system behaviour in the most compact form. If a *state–variable* v_j is partially covered and the set Le_j only contains one element, then v_j can take only two values, either $\infty_j^{Le_j}$ or the value in Le_j. If an *a–machine* is constructed with only this *state–variable*, it will contain two *states* and is easy to see that it is a Boolean lattice. In one *state*, the glb, the *state–variable* is refined and the status of the system is known. In the other *state*, the lub, the *state–variable* can take any value except the one in the glb. It can be easily seen that each *state* is a complement of the other.

Two partially covered *state–variables* of the type mentioned in the previous paragraph generate a Boolean lattice with 2^2 *states*. Examples are the three *a–machines* that describe the functional behaviour in the buffer system example of section 2.7 (figs. 2.10, 2.11 and 2.12). Each of these *a–machines* contain three *state–variables* in each *state*. The first *state–variable* (level measurement) is always covered. Therefore, the actual Boolean lattice is constructed with the other two *state–variables*. For instance, in the case of fig. 2.10, the second and third *state–variables* represent the position of the fill and drain valves respectively. In this case, the refined values (the values of interest) occur when valves are closed. These refined values appear in *state* 1 which is the glb of the Boolean lattice. The lub is given by *state* 3. *States* 2 and 4 complete all the possible combinations of open or closed valves. The self–loops in each *state* represent the transitions causing changes in the covered *state–variables*. If the connection between *states* is labelled with the appropriate transitions, this lattice can be interpreted as a model of the system in which a *state–variable* changes from a well identified value to another in which the rest of the possible behaviour is covered. In order to properly consider the execution of the remaining transitions that do not drive the system to the refined value, a "self–loop" is introduced in the *state* containing the partially covered symbol. This self–loop is labelled with all the transitions associated with *state–variable* values different from the one refined (i.e. those transitions either mapping the *state–variable* to a value different from the refined value or mapping the refined value to any other). In order to distinguish this augmented structure from a standard lattice, it will be identified as a *quasi–lattice*.

2.5   Languages and a–machines

The behaviour of a system generated by an *a–machine* is identified as its *language*. Language and Automata Theories give the foundations for a formal and systematic study of discrete behaviour understood as the occurrence of events. The following paragraphs present standard definitions taken from Wonham (1988) concerning languages. A good introduction to Automata and Language Theories is given by Hopcroft and Ullman (1977), while Wonham (1988) presents an overview of its use in Supervisory Control Theory.

A *symbol* is an abstract identity which does not need to be defined. An *alphabet* Σ is a set of symbols. A *string* or *word* is a sequence of symbols. A *prefix* of a string is any number of leading symbols in the string. In this work, symbols are associated with transitions and therefore the alphabet is the set of all possible transitions that can occur in a given process. A string is a sequential occurrence of transitions. In this work, strings will be identified as *trajectories*.

A *language L* is the set of all finite strings that can occur within a string set Σ^*. In process control terms, L is the set of all possible trajectories that can occur in a plant model. In this work, the terms "system behaviour" and language are used to define the same concept. The term language will be used in this chapter and "system behaviour" will be used in later chapters. A *regular language* is a language that can be represented by a *FSM* (Hopcroft and Ullman, 1977):

$$L : \{\sigma \in \Sigma^* \ / \ \delta(\sigma, q_0) \text{ exists}\}$$

Hence, the *a–machine* also models *regular languages*. Here the treatment is restricted to this class of languages.

The *prefix closure*, $\overline{L}$, of a language $L \subseteq \Sigma^*$, is the language which possesses all the prefixes of the strings of L.

$$\overline{L} : \{\sigma \in \Sigma^* \ / \ \exists t \in \Sigma^*, \ \sigma t \in L\}$$

In other words, the prefix closure of a language L realised by a given *FSM* is the set of all partial trajectories that can be constructed from the initial state of the *FSM*, that can be extended to a complete trajectory (e.g. a task) which is part of L.

It is important to note that in the Supervisory Control Theory, for every machine M, $L(M)$ is assumed to be closed. That is $L(M) = \overline{L}(M)$ (Ramadge and Wonham, 1987b, p. 211). Here, this assumption is maintained, thus restricting the *a–machines* to those that are trim.

The *marked language* of a machine is composed of all strings in which the final *state* is a member of the set of marked *states*:

$$L_m(M) : \{\sigma \in L(M) \ / \ \delta(\sigma, q_0) \in Q_m\}$$

It represents all the trajectories finishing at *states* previously defined as relevant for the process operation. With the given definitions it can be said, in terms of languages, that an *a–machine* M is coreachable if any string in $L(M)$ can be completed to a string in $L_m(M)$.

Given $L(M_1)$ and $L(M_2)$, the languages of *a–machines* M_1 and M_2, the language intersection, $L(R) = L(M_1) \cap L(M_2)$, is given by

$$L(R) : \{\sigma \in \Sigma \ / \ \sigma \in L(M_1) \text{ and } \sigma \in L(M_2)\}$$

During the synthesis of a controller, it is very important to guarantee that certain trajectories can always be extended to complete trajectories (for example, the completion of particular tasks). In other words, the language $L(M)$ of the machine M representing the system behaviour must reach the marked *states* of machine M. If this is true, then it is said that M is nonblocking. That is

$$L(M) = \overline{L}_m(M)$$

As can be seen, a nonblocking machine must be reachable and coreachable (i.e. it must be trim).

The same can be said in the case of determining the prefixes and strings which survive when combining languages (Wonham, 1988). Given languages L_1 and $L_2 \subseteq \Sigma^*$, it is always the case that

$$\overline{L_1 \cap L_2} \subseteq \overline{L_1} \cap \overline{L_2}$$

i.e. any prefix common to L_1 and L_2 is a common prefix of L_1 and L_2. However, it cannot be always guaranteed that every common prefix can be completed to a common string. This means that there may exist common partial trajectories in two different systems that cannot be completed to the same complete trajectory. This is critical when dealing with marked *states* guaranteeing the existence of specific subsets of trajectories in different systems. Therefore, the following definition is introduced.

Definition 2.7 *Nonconflicting Languages (Wonham, 1988).*

   Languages L_1 and $L_2 \subseteq \Sigma^*$ are said to be nonconflicting if

$$\overline{L_1 \cap L_2} = \overline{L_1} \cap \overline{L_2}$$

In other words, L_1 does not contain any complete trajectory that is not in L_2 and vice versa. Two languages can only be guaranteed in advance to be nonconflicting if either they are prefix–closed (both contain all possible strings) or their respective alphabets are disjoint (the nonconflicting property is satisfied trivially). Another condition that guarantees nonconflicting is the "nesting" of specifications proposed by Wonham and Ramadge (1988). This is yet to be explored. In the general case, the nonconflicting property must be tested for each pair of languages.

2.6 Operations with a–machines

In section 2.2, a formalism based on Automata Theory was introduced as a modelling tool. All the standard operations for *FSMs* and regular languages can be used within this formalism. Other operators emphasising synchronisation issues are available in the literature (e.g. Heymann (1990)). Here, two basic operators are modified to handle *state–variables* explicitly. First, an interleaved *FSM* product, identified as the *asynchronous product* is defined. The *state–variable* in the resultant *a–machine* is given by the Cartesian Product of the *state–variables* of each *a–machine* in the product. The second operator is the *synchronous product* of *FSMs* extended to handle *state–variables*. It resolves ambiguities introduced by the coverings performing the refinement of *state–variables* using information from the *a–machines* involved in the product. These two operators will be used in the construction of models as will be shown in the following section and in chapters 5 and 6 for the synthesis of the controllers.

2.6.1 Asynchronous Product of a–machines

Section 2.2 alluded to process model construction from elementary *a–machines*. Such elementary *a–machines* describe how each part of the process works (e.g. the positions of an on/off valve and how they change from one position to another). More complex structures can describe relationships among elementary components (e.g. how a level indicator can change in a tank as a function of the position of the fill and drain valves).

One way of constructing an *a–machine* of the overall process is to find all possible combinations of the *state–variables* in an interleaved fashion. This *a–machine* can be constructed using the asynchronous product operator defined as follows.

Definition 2.8 *Asynchronous Product of a–machines* ($\|$).

Given two a–machines $M_1 : \{\sigma \in \Sigma_1; q_1, q_1' \in Q_1\}$ and $M_2 : \{\mu \in \Sigma_2; q_2, q_2' \in Q_2\}$, in which $\Sigma_1 \cap \Sigma_2 = \emptyset$, the asynchronous product $M_R = M_1 \| M_2$ is given by the interleaving of states of each machine

$$M_R \quad : \quad \delta(\sigma, q_R) = q_R'$$
$$\delta(\mu, q_R) = q_R'' \quad and$$
$$\Sigma_R \quad = \quad \Sigma_1 \cup \Sigma_2$$

where the state variables $v_{q_R}, v_{q_R'}, v_{q_R''} \in V_R : V_1 \times V_2$ of the new machine R are given by the Cartesian product of state–variables for each corresponding state:

$$v_{q_R} = (v_{q_1}, v_{q_2})$$
$$v_{q_R'} = (v_{q_1'}, v_{q_2})$$
$$v_{q_R''} = (v_{q_1}, v_{q_2'})$$

$v_{q_1}, v_{q_1'} \in V_1$ and $v_{q_2}, v_{q_2'} \in V_2$

The difference between this and a standard *FSM* product is that the construction of the *state–variable* vector is done explicitly. Examples of the use of this operator are presented in section 2.7.

2.6.2   Synchronous Product of a–machines

During the controller synthesis, or the model construction, it is necessary to calculate the intersection of trajectories among different machines (i.e. the language intersection of the machines involved). For regular languages, this is equivalent to a product in which, from a given *state* of each machine, only identical transitions are considered in the resultant machine. This is named the *synchronous product*. The ambiguities introduced by an *a–machine* due to the symbols ∞_j and $\infty_j^{L_j}$ are resolved using information from any of the *a–machines* being intersected by substituting refined *state–variable* values in the resultant *state* from the intersection. The above is captured in the following definition.

Definition 2.9 *Synchronous Product of a—machines* (§).

Given two a—machines M_1 and M_2, the synchronous product $M_R = M_1 \S M_2$
is given by the language intersection of both a—machines:

$$M_R \quad : \quad \delta(\sigma, q_R) = q'_R \text{ and}$$
$$\Sigma_R \quad = \quad \Sigma_1 \cap \Sigma_2, \ \sigma \in \Sigma_R$$

where $L(R) = L(M_1) \cap L(M_2)$ and $\gamma(\sigma, (v_j)_{q_R})$ takes the refined state—variable
value from $\gamma(\sigma, (v_j)_{q_1})$ and $\gamma(\sigma, (v_j)_{q_2})$.

In other words the *synchronous product* is given by the *a—machine* that
realises the language intersection in which

1. Every trajectory is in $L(M_1)$ and $L(M_2)$.

2. The refined *state—variable* value from *states* in either *a—machines* M_1 or
   M_2 is assigned to the resultant *state—variable* of the product. If different
   refined values are given by each machine, then the product is not defined.

It is important to note that a *FSM* resulting from the *synchronous product*
of two trim *FSMs* is not necessarily trim. For instance, suppose that a marked
state is common to two *FSMs* and is accessed only by two strings (i.e. trajectories) on each machine as shown in fig. 2.1. One of the strings is the same in
each machine but the other is not. Assume that the difference is transition τ_4
missing. Therefore, the marked *state* will not be reached by this sequence and
the resultant *FSM* from the *synchronous product* fails to be coreachable.

2.7 Examples

Two simple examples are used to illustrate the ideas presented in this chapter.
In both cases, models are constructed in a modular and incremental fashion.
Elementary models are defined first for each of the process components and
the overall process model is then constructed incrementally, incorporating each
elementary model using the operators defined in the previous section. In the
first example, the *asynchronous product* is applied to elementary models to
obtain a final structure, which is a Boolean lattice. In the second example,
model construction proceeds by incrementally adding each process component.
Afterwards, the model is simplified, diminishing the number of transitions by
eliminating those transitions without a "physical meaning" (e.g. if the feed
valve is closed, the level will never increase). This is accomplished by introducing additional *a—machines* describing this functional behaviour and performing
the *synchronous product* with the initial model in an incremental basis.

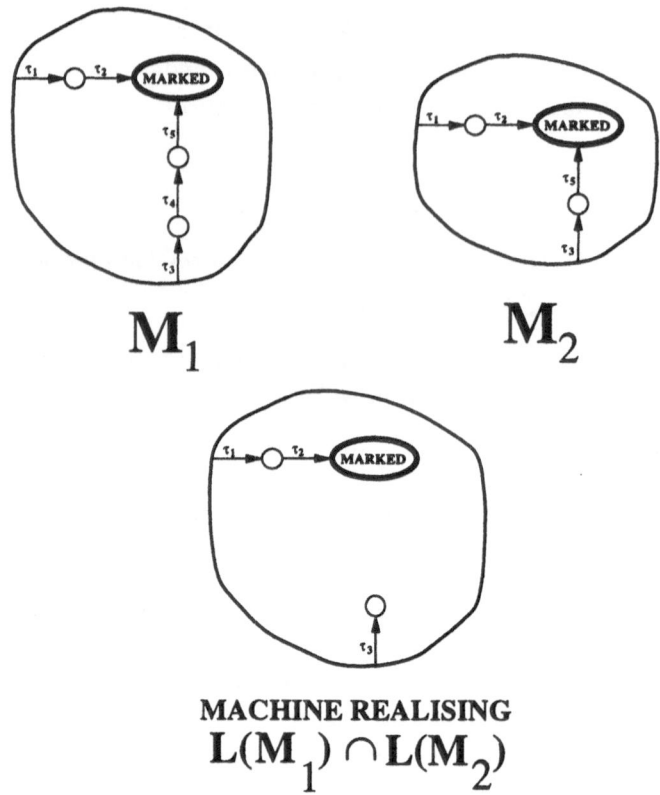

MACHINE REALISING
$$L(M_1) \cap L(M_2)$$

Figure 2.1: A case in which the synchronous product of two trim *FSMs* can generate a non trim *FSM*.

2.7.1 Burner System

The first example is taken from the literature (Moon *et al.*, 1992). The process is the simplified burner system shown in figure 2.2. It consists of two on/off valves to feed air and fuel to a burning chamber, a flame igniter and a flame detector in the chamber. The objective is to generate a model that describes the operation of the overall process. First, elementary models are defined for each process component. Then, the model is obtained as the asynchronous product of all the elementary machines.

2.7.1.1 Elementary Models

The process is divided into the four elementary process components as listed above: air valve, fuel valve, igniter and flame detector. The *a-machines* for each component are presented in table 2.1. Each *a-machine* describes the

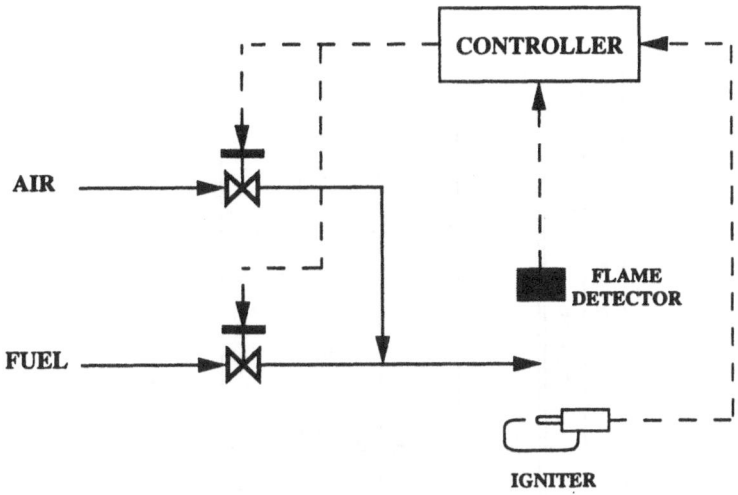

Figure 2.2: The burner system.

operational states and transitions of one process component. For instance, the air and fuel valves, whose *a–machine* model is shown in fig 2.3, are described by two *states* with a *state–variable* value in each *state*. The *state–variable* describes whether the valve is open or closed. The *states* are connected by transitions indicating if the valve is opening (going from closed to open) or closing (going from open to closed). Both transitions are controllable (i.e. by an exogenous signal). Models for the igniter and flame detector are shown in figures 2.4 and 2.5. In this example, transitions 41 and 42, causing the change in the flame detector, are uncontrollable. The detection of the flame is an event that can occur during normal startup as a consequence of other (controllable) transitions or when the burner is on fire caused by a fuel leak or improper operation. In addition, after ignition the flame may go off at any moment, either as a part of the normal shutdown or due to abnormal circumstances in the operation (e.g. fuel line blockage).

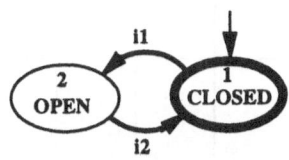

Figure 2.3: *a–machine* M_i corresponding to air and fuel valves, $i = 1, 2$.

elementary component	a-machine	fig.	state-variable			state label	transition			
			description	value	state label	label	description	from st	to st	
air valve	M_1	2.3	valve position	0: closed	1	11	opening	1	2	
				1: open	2	12	closing	2	1	
fuel valve	M_2	2.3	valve position	0: closed	1	21	opening	1	2	
				1: open	2	22	closing	2	1	
igniter valve	M_3	2.4	igniter status	0: off	1	31	switching on	1	2	
				1: on	2	32	switching off	2	1	
flame detector	M_4	2.5	detector status	0: flame detected	1	41*	flame being detected	1	2	
				1: flame not detected	2	42*	flame extinguishing	2	1	

Table 2.1: List of elementary a-machines of the burner system (* = uncontrollable transition).

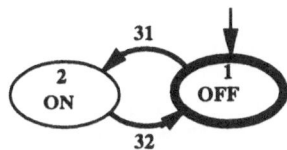

Figure 2.4: *a–machine* M_3 corresponding to igniter.

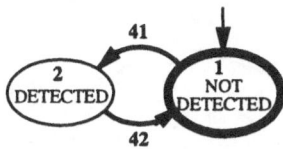

Figure 2.5: *a–machine* M_4 corresponding to flame detector.

2.7.1.2   Process Model

The overall process model is given by the *a–machine* containing the Cartesian product of the set of values of the four elementary *state–variables* obtained by using the *asynchronous product* operator. First, the operator is applied to the *a–machines* modelling the air and fuel valves, obtaining a new *a–machine* which contains all the possible combinations of *state–variable* values from these two *a–machines* (in this case, 4 *states*). The *state–variables* in the resulting *a–machine* take the values of the Cartesian product of the *state–variables* of each elementary *a–machine* (i.e. (air closed, fuel closed), (air open, fuel closed), (air closed, fuel open) and (air open, fuel open)). The next step is to calculate the *asynchronous product* of the resultant *a–machine* and one of the remaining *a–machines*. The process is repeated until all the elementary *a–machines* are included. This generates the set of all possible combinations of *state–variable* values (i.e. the number of *states* in the resultant *a–machine*), with $\prod_{i=1}^{n} m_i$ elements, where m_i is the size of the domain of the *state–variable* associated with elementary *a–machine* i, and n is the total number of elementary *a–machines* in the system.

The four *state–variables* for the overall model are given in the order

(air valve position, fuel valve position, igniter status, flame detector status)

The resultant *a–machine* is shown in figure 2.6. It contains 16 *states*. The initial *state* (indicated by an input arrow) and "normal operation" *state* (label 12) in fig. 2.6 are defined as marked *states*, the first depicting a safe system state or the end of the shutdown procedure while *state* 12 is interpreted as the end of the startup procedure or the normal operating point. In this particular

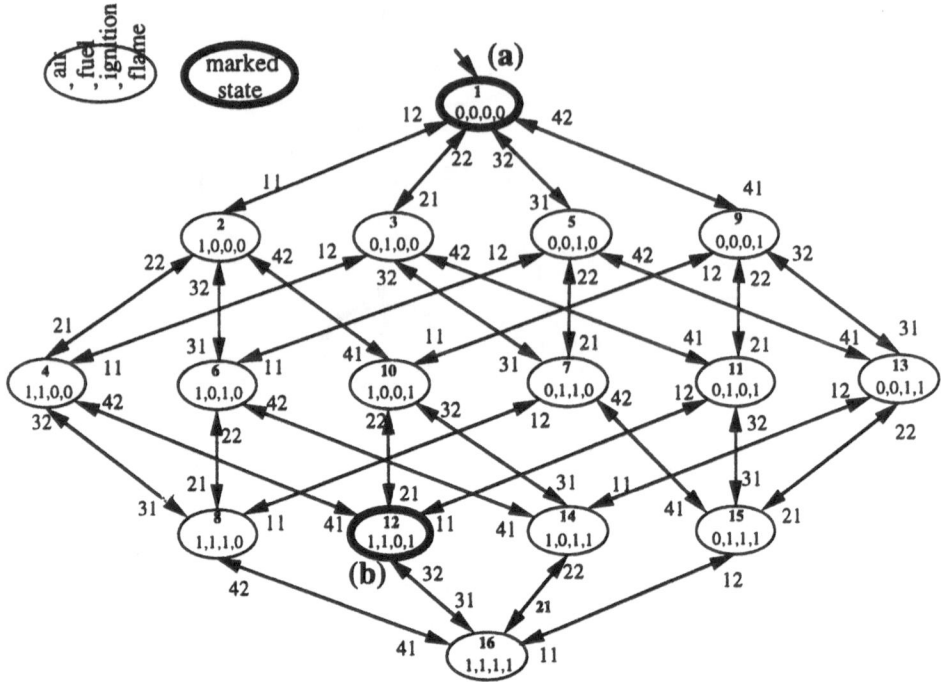

Figure 2.6: Model of the burner system. a = initial/final state; b = normal operation state.

case, it is observed that the *a–machine* is trim and is a Boolean lattice, as can be visually checked. In terms of Language Theory nomenclature, the alphabet is $\Sigma = \{11, 12, 21, 22, 31, 32, 41, 42\}$. A *string (trajectory)* belonging to the *marked* language L_M could be the normal startup procedure given by the following sequence of actions: open air valve, switch igniter on, open fuel valve, detect flame and switch igniter off. This is given by the following sequence of *states* and transitions in fig. 2.6: (*state* 1, transition 11) → (*state* 2, transition 31) → (*state* 6, transition 21) → (*state* 8, transition 41) → (*state* 16, transition 32) → *state* 12. A prefix of the model language L could be, for instance, the initial part of the startup procedure: (*state* 1, transition 11) → (*state* 2, transition 31 → *state* 6.

2.7.2   Buffer System

The second example is the buffer system shown in figure 2.7. It consists of a tank with a level measuring device (i.e. three level switches) to indicate when the level is at its minimum, normal or maximum. Liquid is fed into

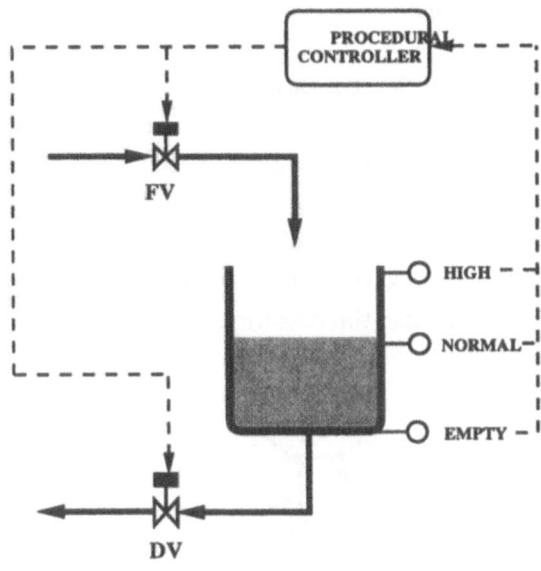

Figure 2.7: The buffer system.

the tank using the on/off valve FV and is discharged through the on/off valve DV at the bottom of the tank. During normal operation, the tank is filled up to the normal level and then emptied. This operation is repeated cyclically. The initial state of the system is when both valves are closed and the level indicator is at its minimum. It is assumed that this is the only state in which the operation can be halted or terminated.

As in the previous example, the objective is to generate a model that describes the operation of the overall process. First, elementary models are defined for each process component. Then, a structure is obtained applying the *asynchronous product* to all the elementary machines. It describes the interleaved behaviour of the process. In order to simplify the model leaving only physically feasible transitions, first relations amongst the elementary components are modelled by *a–machines* by describing the behaviour of the level indicator as a function of valve positions (i.e. input and output flows) assuming a perfect behaviour (i.e. the level indicator and the valves will never fail). Then the *synchronous product* is executed between the interleaved structure previously obtained and these machines. The resultant *a–machine* describes the physical behaviour of the buffer system.

2.7.2.1   Elementary Models

The elementary models for the two on/off valves are constructed as two–*state*
a–machines as in the previous example. The fill valve is named "FV" while
the drain valve is "DV". The level measuring device is described by an *a–*
machine with three *states* and one *state-variable* as shown in fig. 2.8. *State-*
variable values 0, 1 and 2 correspond to minimum, normal and maximum levels
respectively. Transitions between the *states* of the level indicator corresponding
to a change in level are provided in each *state*. The three *a–machines* are
presented in table 2.2. In this example, level indicator transitions 11, 12, 13,
and 14 are uncontrollable because the detection of the changing level depends
only on the valve positions (i.e. input and output flows).

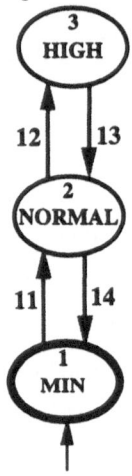

Figure 2.8: *a–machine* M_4 corresponding to the level indicator model.

2.7.2.2   Process Model

First, the Cartesian product of the set of *state-variable* values of the three
elementary system components is obtained using the *asynchronous product* operator following the same procedure as in the previous example. This gives rise
to the *a–machine* with 12 *states* shown in fig 2.9. The order of the elementary
state-variables in each *state* is as follows: (level indicator, FV position, DV
position). This *a–machine* represents all the *states* and their sequencing using the appropriate transitions regardless of their physical feasibility. Marked
states represent conditions in which the system operation can be halted safely.
In this case, the only selected *marked state* is when the tank is empty and
valves are closed (i.e. $Q_m = \{1\}$). In order to simplify the model leaving only

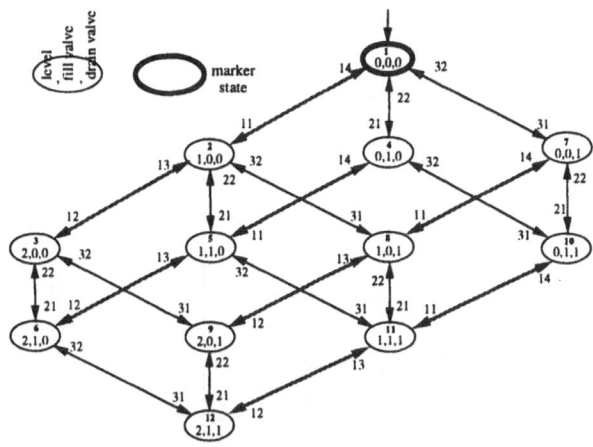

Figure 2.9: Global model of the buffer system.

physically feasible transitions, relations amongst the elementary components are modelled by *a–machines* describing the behaviour of the level indicator as a function of valve positions (i.e. input and output flows). Provided that none of the components fail, the level indicator changes only under specific circumstances given by the position of the two valves. This is described in the following paragraphs.

If both valves are closed, the level indicator must keep the same value until any of the two valves is open. Hence, transitions associated with the level indicator must not occur from this state. This is modelled by the *a–machine* in fig. 2.10. In this *a–machine*, the *state–variable* level indicator is covered and therefore the structure can be considered a Boolean lattice in terms of FV and DV *state–variables*. Therefore the number of *states* is $\prod_{i=2}^{3} |m_i| = 4$. The structure matches the definition of quasi–lattice introduced in subsection 2.4 because of the presence of self–loops in each *state* corresponding to the covered *state–variable* level indicator. The initial *state* in the *a–machine*, indicated by an input arrow, corresponds to both valves closed. This is the lattice glb. From this *state* the only transitions leading to other *states* are the opening of valves. Transition 21 to *state* 2 changes the FV position from "closed" to anything else except closed. In this case, "open" is the only possible value for this *state–variable* but the partial covering symbol $\infty_2^{Le_2}$, $Le_2 = \emptyset$, is used in order to maintain the generality of the representation. *State* 4 is generated in the same way. In these *states*, all the transitions related to a change in level are permitted via the self–loop. The transitions going out from these *states* are only those

elementary component	a–machine	fig.	state–variable description	value	state label	transition label	description	from st	to st
level indicator	M_5	2.8	level status	0: min	1	11*	from min to normal	1	2
				1: normal	2	12*	from normal to max	2	3
				2: max	3	13*	from max to normal	3	2
						14*	from normal to min	2	1
fill valve FV	M_1	2.3	valve position	0: closed	1	21	opening	1	2
				1: open	2	22	closing	2	1
drain valve DV	M_1	2.3	valve position	0: closed	1	31	opening	1	2
				1: open	2	32	closing	2	1

Table 2.2: List of the elementary *a–machines* of the buffer system (* = uncontrollable transition).

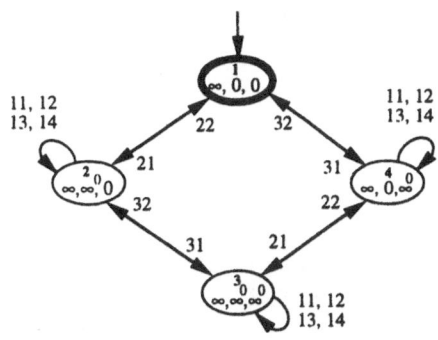

Figure 2.10: *a–machine* prescribing functional behaviour of the buffer system when both valves are closed.

related to opening or closing FV or DV. The lub contains the level indicator *state–variable* covered and the valve position *state–variables* partially covered. The connections are done as in *states* 2 and 4. The only marked *state* is the initial *state*. It is easy to see that when performing the *synchronous product* between the *a–machine* in fig. 2.9 and this quasi–lattice, level transitions will be eliminated from *states* in which both valves are closed. Only transitions related to the opening of the valves will be retained.

Following the same procedure, similar *a–machines* are constructed to prescribe the behaviour when only one valve is open. If FV is open and DV is closed, the level will continuously increase until the maximum level is reached or until FV is closed or DV is open. The *a–machine* in fig. 2.11 represents such behaviour. As in the previous case, the only marked *state* is that in which both valves are closed. Note that in this case the lattice glb is given by *state* 2 and it does not correspond to the initial *a–machine state* indicated by the input arrow.

On the other hand, if DV is open and FV is closed the level will decrease until the tank is empty, or either FV opens or DV closes. Such behaviour is captured in the *a–machine* of fig. 2.12. The previous comments regarding marked and initial *states* apply. In the present case the glb is *state* 2 while lub is *state* 4.

The *synchronous product* among these three *a–machines* and the *a–machine* of fig. 2.9 is obtained and shown in figure 2.13. Compared to fig. 2.9, it is observed that the role of the *a–machines* describing the functional behaviour of the tank is to eliminate some transitions among *states* and to leave only those that are feasible according to the described behaviour. The only marked *state* is the initial *state*.

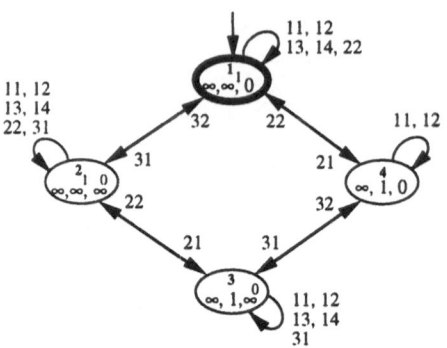

Figure 2.11: *a-machine* prescribing functional behaviour when FV is open and DV is closed.

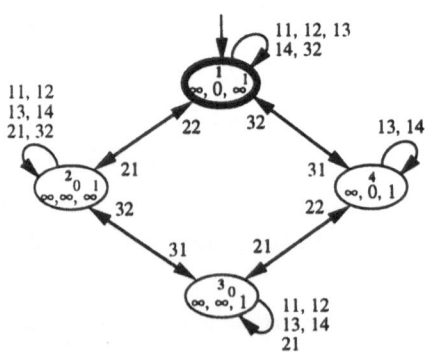

Figure 2.12: *a-machine* prescribing functional behaviour when DV is open and FV is closed.

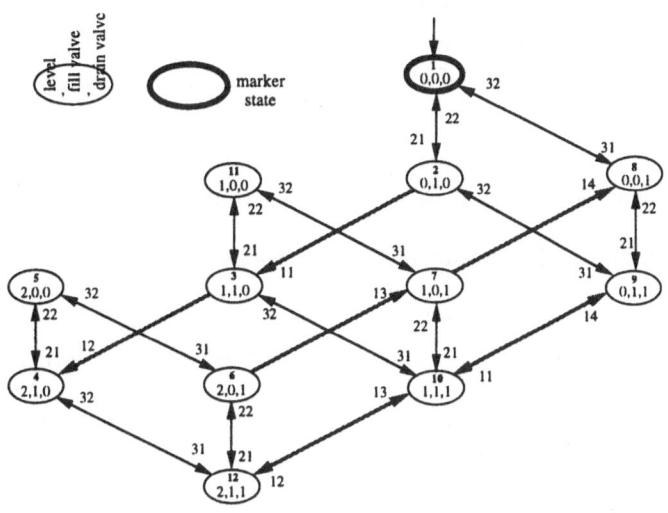

Figure 2.13: Model for the buffer system.

2.8   Summary

A labelled *FSM*, named *a-machine*, is proposed as a basic modelling tool in which each *state* is characterised by a set of *state-variables* describing the process. *State-variables* take values from finite domains. Two extra symbols are included to handle incomplete information in the model and to facilitate its construction: the covering (∞_j) and the partial covering $(\infty_j^{Le_j})$ symbols. The covering symbol can be assigned to any *state-variable j* for which no information is available or if it is of no interest to the current *state*. The partial covering symbol excludes from the possible values of *state-variable j* those defined in the subset Le_j. Partial orders on the *states* and *state-variables* are introduced to formally handle these extra symbols and to compare *state-variables* and *states*. They also introduce an ordering in the *states* resulting in compact and well defined structures from an algebraic point of view. As it will be shown in the next chapter, they also establish the formal link for the mapping of behaviour specifications modelled by logic formalisms into the *FSM* domain in which the controller synthesis is performed.

In section 2.5, general definitions from Language Theory were introduced. These will be used in chapter 4 when dealing with the Supervisory Control Theory.

Since *a-machines* are *FSMs*, operators from Automata Theory can also be used for their manipulation. Section 2.6 introduced two of these operators, modified so as to consider explicitly the handling of *state–variables*. These are the *asynchronous product* and the *synchronous product*. Their use is demonstrated in the construction of models. They will also be employed for controller synthesis.

Finally, in section 2.7, two examples demonstrate the model construction using the ideas presented in this chapter, including:

- Modelling of elementary process components as *a–machines*.

- The use of covering and partial covering symbols to handle incomplete information and to maintain a compact model representation.

- Definition of partial order on the *state–variables* and *states* in order to generate a structured model representation.

- Construction of the process model in an incremental fashion using the proposed operators.

Although state explosion is still an issue, the use of structured models with defined algebraic operations allows the handling of process behaviour by representing explicitly a relatively small number of states. This is particularly useful in the construction of behaviour specification models, as will be shown in the following chapter. Also, it opens the possibility for efficient computer implementations of the required operations. However state explosion is still a subject for further work.

Chapter 3

Specification Modelling

3.1 Introduction

The synthesis of supervisors and controllers, as will be discussed in chapter 5, requires 1) a model of the process behaviour, 2) a model of the desired behaviour to be imposed on the process. Efficient ways of dealing with behaviour specifications related to static specifications (logic invariant properties) have been proposed (Ramadge and Wonham, 1987a; Kumar *et al.*, 1991b). However no methods have been developed for the construction of specification models considering dynamic behaviour (dynamic specifications). Our previous experience has shown that building such models is extremely difficult even for simple specifications. Similar experiences have been reported in the literature (Brooks *et al.*, 1990).

In this chapter, the prescriptive power of logic formalisms is used to overcome the difficulties in constructing behaviour specifications. They are classified according to the formalisms employed to model them. Specification related to logic invariant properties, such as limits on buffer capacities or avoidance of forbidden states, are identified as static specifications and modelled using Predicate Logic statements. In this work, the treatment is restricted only to forbidden states. Specifications describing dynamic behaviour are modelled by TL formalisms. The prescriptive power and formal syntax of these formalisms are exploited so that the specification task proceeds in a modular fashion. Section 3.2 discusses briefly the relationship between *FSMs* and Predicate Logic (PL) statements as presented by Ramadge and Wonham (1987a) and Kumar *et al.* (1991b). The use of covering and partial covering symbols to represent forbidden states is then explained. Section 3.3 deals with the modelling of dynamic specifications using a restricted type of (linear) Temporal Logic (LTL) formulas and its translation into the *a–machine* domain. First, general concepts are introduced and then the syntax of the restricted type of LTL formulas is discussed. These formulas are characterised in terms of LTL subformulas called schemes. The following subsection presents homomorphisms to translate the schemes into the *a–machine* domain together with a translation procedure. The structure of lattices is exploited to generate compact representations. Finally, in section 3.4, the burner example is used to illustrate the modelling of forbidden states and dynamic specifications and their translation into the *a–machine* domain. The specifications cover the startup, normal operation and shutdown procedures. They prescribe the desired behaviour based on the sequencing of events, such as a process startup, as well as specifications to avoid undesirable behaviour, such as the reaching of dangerous states. The obtained *states* corresponding to the forbidden states and the *a–machines* corresponding to the dynamic specifications, together with the model developed in chapter 2 will be used in chapter 5 to synthesise a suitable procedural controller for the burner.

3.2   Forbidden States Modelling

The use of predicate statements as static specifications has been studied in the past by Ramadge and Wonham (1987a) and Kumar *et al.* (1991b). Predicates are associated with *states* in Q and it is shown that there exists an isomorphism h between the family of all predicates $p \in \mathcal{P}$ associated with Q and the Boolean lattice of the power set of Q under the correspondence

$$h : \quad Q \leftrightarrow \mathcal{P} \ \{q : q \in Q \text{ and } p(q) = 1\}$$

Therefore the terms "state" or "predicate statement" can be considered essentially the same in this context. Also, Kumar *et al.* (1991b) showed how an *FSM* can be represented in terms of predicate transformers (Dijkstra and Scholten, 1990) making concepts from the Supervisory Control Theory, such as controllability, amenable to expression in terms of predicate statements. Here, these ideas are not explored further and the treatment is limited to modelling forbidden states in the system.

The standard symbols of Predicate Logic used here are given in table 3.1. Note that the exclusive disjunction symbol is included. It will be of particular use when constructing predicates describing the sequential occurrence of events.

Given the existence of isomorphism $h : Q \leftrightarrow \mathcal{P}$ and homomorphism $\beta : Q \rightarrow V^{n_v}$ defined in section 2.3, it is possible to represent in terms of predicates the

Connectives	:	$\neg$ (negation) , $\rightarrow$ (implication), $\wedge$ (conjunction), $\vee$ (disjunction), ∇ (exclusive disjunction)
Parenthesis	:	$\{, [, (,) ,], \}$
Predicate statements called atomic formulas		

Table 3.1: List of Standard Predicate Logic symbols used in this work

state-variables corresponding to its isomorphic *state* such that

$$H(p) = \beta(h^{-1}(p)) = (v_j)_{h^{-1}(p)}, \ \ p \in \mathcal{P}, \ \ j = 1 \ldots n_v$$

This suggests the construction of a *state* in the Predicate Logic domain as the conjunction of atomic formulas representing the different *state-variables* of the given *state*:

$$h(q) = a_1 \wedge a_2 \wedge \ldots \wedge a_{n_v}$$

where a_1, a_2, ..., a_{n_v} are atomic formulas representing the *state-variables*. For the rest of this work it will be assumed that an atomic formula symbolised by a_j refers to a *state-variable* v_j. For simplicity of notation, letter p will be used to represent those predicate statements of the form ($a_1 \wedge a_2 \wedge \ldots \wedge a_{n_v}$), which can be translated into *state-variables* of a given *state* through homomorphism H.

The symbol ∞_j as a possible value of *state-variable* v_j is translated as a propositional statement formed by the exclusive disjunction of all the values covered by ∞_j. This means that the instantiation of a predicate statement representing a *state* into its equivalent propositional statement may give rise to an expression representing several *states* in an *a-machine*. For instance, assume the existence of a *state* $h^{-1}(p)$ with three *state-variables*, $p = (a_1 \wedge a_2 \wedge a_3)$ in which each of the atomic propositions can be instantiated to values s_{i1}, s_{i2}, s_{i3} as well as ∞_i, $i = 1, 2, 3$ (*state-variable* number). Assume that a_1 is instantiated to value $\infty_1 = s_{11} \nabla s_{12} \nabla s_{13}$, while a_2 and a_3 are instantiated to values s_{21} and s_{32} respectively. This gives rise to the following propositional statement:

$$p = ([s_{11} \nabla s_{12} \nabla s_{13}] \wedge s_{21} \wedge s_{32})$$

which can be distributed to represent three states, as follows:

$$p = (s_{11} \wedge s_{21} \wedge s_{32}) \nabla (s_{12} \wedge s_{21} \wedge s_{32}) \nabla (s_{13} \wedge s_{21} \wedge s_{32})$$

For a *state-variable* v_j taking the value $\infty_j^{Le_j}$, there is a variable $v_k = Le_j$ that represents its complement. In the Propositional Logic domain, ∞_j^{Lej} is interpreted as the negation of elements in Le_j in disjunctive form in a domain in which only sequencing is allowed. For instance, if $a_1 = \infty_1^{Le_1}$ in the statement from the previous example, then *state* $q = (\infty_1^{Le_1}, v_2, v_3)$ with $Le_j = s_{11}$ generates the proposition $p_a = (\neg[s_{11}] \wedge s_{21} \wedge s_{32})$ which is equivalent in logic terms to $p_b = ([s_{12} \; \triangledown \; s_{13}] \wedge s_{21} \wedge s_{32}) = (s_{12} \wedge s_{21} \wedge s_{32}) \; \triangledown \; (s_{13} \wedge s_{21} \wedge a_{32})$. Forbidden states are specified as the negation of a propositional statement of this type. Examples are presented in section 3.4.

3.3   Dynamic Specifications Modelling

Quantitative time is not treated by the modelling structures presented in chapter 2. The passage of time is associated with the function δ that imposes an order of execution on the *a-machine states*. Therefore, dynamic behaviour specifications deal only with this type of qualitative–time characteristic, such as the sequencing of events or the eventual execution of a specific action. Temporal Logic (TL) is a modal logic in which its operators are interpreted in a qualitative time domain (i.e. the accessibility relation is interpreted as the passage of time). As mentioned in chapter 1, time imposes the main standard of classification of existing TL frameworks. The two main versions are linear or branching. In linear time Temporal Logic (LTL), formulas are interpreted as linear sequences, while in branching time Temporal Logic (BTL), trajectories are allowed to branch. The pros and cons of these approaches can be found in the literature (Emerson and Halpern, 1986; Galton, 1989; Ostroff, 1989b). In interpreting state–transition structures, BTL has more descriptive power due to its ability to treat properties occurring in different branches whereas LTL can describe properties occurring in only one branch. LTL may also introduce ambiguities when using negations (Emerson and Halpern, 1986). However, BTL requires a larger amount of TL operators. In the present work, a restricted syntax taken from the framework devised by Ostroff (1989b) and Thistle and Wonham (1986) based on Manna–Pnueli's LTL system (Manna and Pnueli, 1982) has been chosen on an experimental basis. Ostroff's framework, among other characteristics, uses standard predicate statements plus an extra predicate, τ called "next transition" as the basic building blocks. Predicate τ considers the execution of a transition as a statement that can be included in a LTL formula regardless of the use of the temporal operators. The other main advantage, which has not been exploited here, is that the axiomatic system on which the framework is based has been adapted to consider real–time systems and not only "computer programs" as was Manna and Pnueli's. Moreover, the associated proof system might be used to simplify complex formulas in terms of LTL operators and structures and obtain equivalent simpler LTL formulas that, as will be shown in the following subsections, can be matched with pre-defined syntactic structures and then translated into the *a–machine* domain. For simplicity of notation, in the rest of this work TL will stand for linear time Temporal Logic.

In the following paragraphs, the restricted syntax of the TL formulas used here is introduced as well as the way in which these formulas are interpreted in the TL framework. Only three TL operators are included: next ($\bigcirc$), eventually ($\diamondsuit$) and always ($\square$). Other operators such as until ($\mathcal{U}$), unless (U) or precedes (Π) can be derived from these three (Ostroff, 1989b). Formulas are classified into families with the same syntactic structures for which homomorphisms into the *a–machine* domain are developed. The idea is to construct TL models (in the *a–machine* domain) from formulas that are "valid" in a TL frame representing the process behaviour. The construction is made using elementary process component models as defined in chapter 2 guaranteeing any formula to be true if its translation into the *a–machine* domain exists. The translation procedure is later shown in detail.

3.3.1 Structure (syntax) of TL

The basic components used to construct the restricted type of TL formulas are the following:

3.3.1.1 Symbols and Operators

- Standard Symbols of Predicate Logic as listed in table 3.1

- Temporal Logic Operators (TLOp)

  - Next ($\bigcirc$)
  - Eventually ($\diamondsuit$)
  - Always ($\square$)

- "Next transition" statement (τ)

3.3.1.2 Terms

The terms are defined intuitively as follows:

- All predicate statements are terms and are identified as atomic formulas. The set Φ is the set of all atomic formulas.

- If $t_1, t_2, \ldots, t_n$ are terms and f is a n-ary function, then $f(t_1, t_2, \ldots, t_n)$ is also a term.

- No other strings of symbols are terms.

3.3.1.3 Formulas

A formula $A \in \mathrm{Fma}(\Phi)$ where $\mathrm{Fma}(\Phi)$ is the set of all formulas derived from Φ, is defined as an expression constructed from atomic formulas with predicate symbols, function symbols and TL operators. A distinction between two types of formulas is made, formula and state–formula. A formula will be understood as either a state–formula f or a TLOp f where TLOp is a temporal operator. A state–formula is any formula not containing any temporal operators.

If A is a formula then $\bigcirc A$, $\Diamond A$ and $\Box A$ are formulas. No other strings or symbols are formulas.

The terms are used to name objects of the universe, whereas the formulas make assertions about the objects.

A collection of formulas having a common syntactic form is called a scheme.

The formulas used in this work must be constructed using the following schemes, expressed in BNF notation:

$$[p \mid p \wedge \tau] \to \bigcirc[p \mid (\overline{\bigvee}\tau)] \tag{3.1}$$

$$[p \mid p \wedge \tau] \to \Box[p \mid A] \tag{3.2}$$

$$[p \mid p \wedge \tau] \to \Diamond[p \mid A] \tag{3.3}$$

These schemes mean that formulas must be written with the first element either a predicate representing a *state* or a predicate and a "next transition" symbol. Then the implication symbol must be added followed by any of the TL operators defined previously. For the TL operator next ($\bigcirc$), the next element in the formula is either a predicate describing a *state* or an exclusive disjunction of "next transition" symbols describing the following actions to be executed. In the case of always ($\Box$) and eventual ($\Diamond$) operators, the valid elements are either a predicate describing a *state* or another formula.

3.3.1.4   Frames and Models

A frame is a pair $F = (S, R)$, where S is a non–empty set and $R : S \times S$ is a binary relation.

A model is a triple $M = (S, R, V)$, with $V : \Phi \to 2^S$. The function V assigns to each atomic formula $a \in \Phi$ a subset $V(a)$ of S. In other words, $V(a)$ defines the elements of S in which a is true.

A formula A is true in model M, denoted $M \models A$, if it is true at all elements in M, i.e. if

$$M \models_s A, \quad \forall s \in S$$

A formula A is valid in frame $F = (S, R)$, denoted $F \models A$, if

$$F \models_M A, \quad \forall M = (S, R, V) \text{ based on } F$$

In other words, a valid formula is a formula which is true in all elements of every model.

A scheme is said to be valid in a frame (respectively, true in a model) if all instances of the scheme are valid (respectively, true). Then, it is said that the frame validates the scheme.

3.3.2   Translation of TL Formulas into a–machines

The translation of TL formulas into *a–machines* is based on homomorphisms for each of the schemes mentioned above. These homomorphisms define sufficient conditions to guarantee that the resultant *a–machine* is the TL model in which the TL formula holds and is a subset of a TL frame in which the TL formula is valid. A TL frame for a given process is provided by the *asynchronous product* of all elementary process components (i.e. all the possible interleaved behaviour that the elementary components can generate). Having developed this frame, a TL model (i.e. an *a–machine*) is constructed for each dynamic specification modelled by the three schemes described above and their combination. Schemes are combined by formula A appearing in schemes 3.3 and 3.2. This formula can be constructed using any of the three schemes. The translation into the *a–machine* domain is executed with morphisms for each one of the schemes using the same elementary process components utilised in the construction of the TL frame. Consequently, if the behaviour prescribed in the TL formula is not representable by the combination of the behaviour of elementary process components, then the formula is not true and therefore not valid. For instance, consider the case of the two on/off valves of the burner system modelled in subsection 2.7.1. The *asynchronous product* of the *a–machines* modelling these two valves generates the *a–machine* shown in fig. 3.1. Each of the four *states* contains *state–variable* values in the order (air valve position, fuel valve position). The description of the transitions which open and close each valve are given in table 2.1. This *a–machine* describing the behaviour of the two valves can be used as a frame in the TL domain. Any specification to be true, must consider the behaviour modelled by this *a–machine*. For example, the specification $(1,1) \rightarrow \bigcirc(\tau = 11)$ (i.e. if the air and fuel valves are open, then the next action is to open air valve) gives rise to a TL model (an *a–machine*) that first of all, could not be constructed with the behaviour of the elementary process components provided and secondly, is obviously not true and therefore not valid in the TL frame.

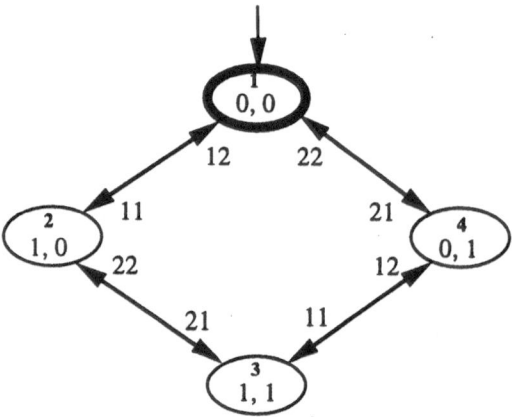

Figure 3.1: *a–machine* corresponding to the *asynchronous product* of the *a–machine* models of the air and fuel valves.

The first step in the translation of a given TL formula is to identify or create the appropriate *state* in the *a–machine* domain from where to start the translation. Such a *state* must contain the same *state–variable* values as the first state of the TL formula under consideration. This is accomplished by constructing a quasi–lattice in which the glb is a *state* containing the initial *state–variable* values given by the TL formula and the lub is its complement. As was shown in section 2.4, this structure contains all the system behaviour described by the set of elementary process models. Once the initial state of the TL formula under translation has been created, the rest of the TL formula is constructed using the appropriate morphisms. At each step of the translation procedure, the validity of the TL formula is established by checking that each transition leads to the appropriate *state–variable* values using the elementary *a–machine* process models.

3.3.2.1   Generation of Initial State in the a–machine Domain

A TL formula may start with a specific state which does not correspond to
the actual initial *state* in the system (e.g a *state* which is the result of the
sequential occurrence of different transitions). This state may contain refined
values or (partially) covered values. It must appear within the *a–machine* under
construction in such a way that the behaviour specified by the TL formula
being translated is shown explicitly in the *a–machine* domain while the rest
of the possible system behaviour appears in the most compact possible form.
In order to achieve this, exploiting the interleaving nature of the *a–machine*
structure, a Boolean lattice is constructed in which the glb *state–variable* values
correspond to those in the TL formula state and the lub corresponds to its
complement. The rest of the structure is constructed using the partial covering
given in definition 2.6, which orders the power set of *state–variable* values in
the Boolean lattice. Transitions taken from the elementary process models are
used to label the transitions connecting each *state* in the lattice. For a given
state–variable in a *state*, the connections in the direction of the lattice glb
(from partially covered to refined) are labelled by all the transitions defined
in the elementary *a–machine* corresponding to the *state–variable* leading to
the specified refined value. In the direction of the lub, the connection is any
transition that goes from the given refined *state–variable* value to any other
value in the corresponding elementary *a–machine*.

Once the Boolean lattice is constructed, self–loops are added to each *state*
containing those transitions associated with values of the covered or partially
covered *state–variables* which are not used as connections within the structure.
This quasi–lattice contains all the possible behaviour from "almost any state"
(the partially covered *state*), to the state appearing first in the TL formula
being translated. For the case of a quasi–lattice having a totally covered initial
state, the structure reduces to only that *state* and its corresponding self–loops.
If the initial state in the TL contains negation statements, the construction
of the *a–machine* is done by substituting all the negations into affirmative
statements and then proceeding as above. Afterwards, a search in the quasi–
lattice is performed to locate the *state* containing the negations as well as it
complement. These two *states* are then defined as the glb and lub respectively.

Once the first state defined by the TL formula has been located in an
a–machine containing all the system behaviour, the translation of the rest
of the TL formula is done using homomorphisms defined for each one of the
three schemes introduced in section 3.3.1. Each scheme generates four different
structures. In the following paragraphs, mappings for these structures are

presented.

3.3.2.2   Translation of Schemes involving TL Operator ◯

In the TL scheme involving TL operator ◯, the same meaning can be assigned to TL operator ◯ or construction $p \wedge (\overline{\vee}\tau)$ as for function δ in the $a\text{-}machine$ domain in which $|\sigma| = 1$. That is, the occurrence of one $state$ after another such that $\delta : q \overset{\delta_s}{\rightarrow} q'$. Moreover, isomorphism $h : q \leftrightarrow p$ defined in section 3.2, relates the $a\text{-}machine$ $states$ with statements describing states in the TL domain allowing the construction of state–transition structures in the $a\text{-}machine$ domain. According to the scheme shown in equation 3.1, TL operator ◯ can be used in four different cases as described in the following paragraphs:

1. The RHS of the implication symbol models only a state (i.e. $\ldots \rightarrow \bigcirc p'$). Regarding the LHS, two different situations arise:

   (a) A "next transition" is found together with the predicate describing the state p (i.e. $p \wedge \tau \rightarrow \bigcirc \ldots$). Given this information and using function δ in the appropriate elementary models, it is possible to determine the next state given the current state and the "next transition". This must correspond to the state given in the RHS of the scheme, if any. Note that the existence of other transitions for which function δ is defined in the current state is not overruled. For this scheme to be true, an $a\text{-}machine$ $m_{p \wedge \tau \rightarrow \bigcirc \ldots} = (Q, \Sigma, V^{n_v}, \delta, \gamma, q_0, v_0)$ must exist such that

   $$\delta(\sigma, h^{-1}(p)) \text{ is defined at least for } \tau = \sigma, \ \sigma \in \Sigma$$

   (b) No transitions accompany the predicate describing the state p (i.e. $p \rightarrow \bigcirc \ldots$). In order for this scheme to be true, the next state p' to be reached must be determined with the scheme RHS by the execution of only one transition. In other words, if the scheme is true, there exists an $a\text{-}machine$ $m_{p \rightarrow \bigcirc p'} = (Q, \Sigma, V^{n_v}, \delta, \gamma, q_0, v_0)$ such that

   $$\delta(\sigma, h^{-1}(p)) = h^{-1}(p') \text{ is defined for } \sigma \in \Sigma \text{ such that } |\sigma| = 1$$

2. The RHS of the implication symbol models the execution of (exclusive disjunction of) "next transitions", $(\ldots \rightarrow \bigcirc \overline{\vee}\tau)$. Regarding the LHS, again two situations can occur.

   (a) No transitions accompany the predicate describing the state p (i.e. $p \rightarrow \bigcirc \ldots$). This means that the list of executable transitions from state p are given by the (exclusive disjunction of) "next transitions", $(\overline{\bigvee \tau})$. For this scheme to be true, an *a–machine* $m_{p \to \bigcirc (\overline{\bigvee \tau})} = (Q, \Sigma, V^{n_v}, \delta, \gamma, q_0, v_0)$ must exist such that

$$\delta(\sigma, h^{-1}(p)) \text{ is defined for all } \sigma \text{ in } \left(\bigvee \tau\right)$$

(b) A "next transition" is found together with the predicate describing the state p (i.e. $p \wedge \tau \to \bigcirc \ldots$). This is the combination of previous cases 1a and 2a. Given this information and using function δ of the elementary components, the next state $h^{-1}(p')$ obtained from the current state $h^{-1}(p)$ and "next transition" τ is calculated as in case 1a. From this next state $h^{-1}(p')$, all the transitions given in $(\overline{\bigvee \tau})$ must be executed as in case 2a. For this scheme to be true, an *a–machine* $m_{p \wedge \tau \to \bigcirc (\overline{\bigvee \tau})} = (Q, \Sigma, V^{n_v}, \delta, \gamma, q_0, v_0)$ must exist such that

$$\delta(\sigma, h^{-1}(p)) \text{ is defined at least for } \tau = \sigma, \ \sigma \in \Sigma$$

$$\text{If } \tau = \sigma, \text{ then } \delta(\sigma, h^{-1}(p)) = h^{-1}(p')$$

$$\delta(\sigma, h^{-1}(p')) \text{ is defined for all } \sigma \text{ in } \left(\bigvee \tau\right)$$

The *states* and *state–variables* in all the *a–machines* defined above are given by

$$H(p) = (v_j)_{h^{-1}(p)}, \ j = 1 \ldots n_v$$

$$H(p') = (v_j)_{h^{-1}(p')}, \ j = 1 \ldots n_v$$

$$h^{-1}(p), \ h^{-1}(p') \in Q$$

$$(v_j)_{h^{-1}(p)}, \ (v_j)_{h^{-1}(p')} \in V^{n_v}$$

A scheme with the TL operator $\bigcirc$ does not define by itself termination conditions for the translation process, as for instance, is the case for TL operator $\square$. It only describes the next *state* to be reached. Thus the *a–machine* representing this behaviour must consider the possibility of the execution of further actions. Therefore, if scheme $[p \mid p \wedge \tau] \to \bigcirc p'$ occurs in the most RHS of a given TL formula in which p' is the final predicate describing a state, the values for the destination *state–variable* $(v_j)_{h^{-1}(p')}, j = 1 \ldots n_v$ are generated from the *state–variable* values in the current *state* $h^{-1}(p)$ using the mapping outlined above. Then, a search is made within the *a–machine* to find a *state* $h^{-1}(p'')$, different from $h^{-1}(p)$, that contains the largest number of *state–variable* values equal to the values of $(v_j)_{h^{-1}(p')}$ with the remaining values being covered or partially covered. The connection between $h^{-1}(p)$ and $h^{-1}(p'')$ is done using the transition that makes the appropriate scheme become true. If *state* q'' is a (partially) covered *state*, it means that no behaviour explicitly exists on the *a–machine* related to this TL scheme. If the transition that makes the scheme true is part of a self–loop in an elementary *a–machine*, then the state must be connected to itself. In the case that two partially covered *states* exist with the same *state–variable* values as $h^{-1}(p'')$, then it is very probable that a glb for the two *states* exist. If the *a–machine* is a quasi–lattice, the glb will exist and it will be found during the search.

3.3.2.3   Translation of Schemes involving TL Operator □

For schemes involving the TL operator □, the objective is to generate models in which this scheme holds except in the trivial case (when both sides of the implication are false). The structure defined in scheme 3.2 gives rise to four arrangements as in the previous case. The description of these four arrangements follows.

1. The RHS of the implication reduces to a state–formula p' representing a *state* in the *a–machine* domain $(\ldots \to \Box p')$. In this case, the LHS can represent a state with an associated transition or only a state in the *a–machine* domain.

   (a) If the LHS of the scheme defines a state and a "next transition" (i.e. $p \wedge \tau \to \Box \ldots$), a model in which this scheme holds is given by *a–machine* $m_{p \wedge \tau \to \Box p'} = (Q, \Sigma, V^{n_v}, \delta, \gamma, q_0, v_0)$ where

   $\delta(\sigma, h^{-1}(p))$ is defined at least for $\tau = \sigma$, $\sigma \in \Sigma$

   If $\tau = \sigma$, then $\{\delta(\sigma, h^{-1}(p)) = h^{-1}(p')$;

   $\gamma(\sigma, \beta[h^{-1}(p)]) = (v_j)_{h^{-1}(p')}, j = 1 \ldots n_v$

and

   $\forall \sigma, \ \sigma \neq \tau$ and $\gamma(\sigma, \beta[h^{-1}(p')])$ being covered or partially covered, then

   $\delta(\sigma, h^{-1}(p')) = h^{-1}(p')$

   (b) If the LHS of the implication only defines a *state*, a model for this scheme is similar to the one described above but in which

   $\delta(\sigma, h^{-1}(p))$ exists for $\sigma \in \Sigma$ such that $|\sigma| = 1$

2. For the scheme in which the RHS of the implication is modelled by formula A, a model must be generated independently for the formula using
   the appropriate morphism.

   (a) If the LHS of the scheme defines the output state and a transition
       (i.e. $p \wedge \tau \rightarrow \Box \ldots$), a model in which this scheme holds is given by
       a–machine $m_{p \wedge \tau \rightarrow \Box A} = (Q, V^{n_v}, \delta, \gamma, q_0, v_0)$ where

       $\delta(\sigma, h^{-1}(p))$ is defined at least for $\tau = \sigma, \sigma \in \Sigma$

       If $\tau = \sigma$, then $\{\delta(\sigma, h^{-1}(p)) = h^{-1}(p')$;

       $\gamma(\sigma, \beta[h^{-1}(p)]) = (v_j)_{h^{-1}(p')}, j = 1 \ldots n_v$

       in which $h^{-1}(p)$ is the final state in the scheme LHS and $h^{-1}(p')$ is
       the initial state in scheme RHS.

       Note that $h^{-1}(p)$ can be a final *state* in which the refined *state-
       variables* do not change in value and there exists a self–loop labelled
       by all the transitions related to the covered or partially covered
       state-variables.

   (b) If the LHS of the implication only defines a state, a model for this
       scheme is similar to the one described above but in which

       $\delta(\sigma, h^{-1}(p))$ exists only for $\sigma \in \Sigma$ such that $|\sigma| = 1$

       $\delta(\sigma, h^{-1}(p)) = h^{-1}(p')$

       $\gamma(\sigma, \beta[h^{-1}(p)]) = (v_j)_{h^{-1}(p')}, j = 1 \ldots n_v$

       in which $h^{-1}(p)$ and $h^{-1}(p')$ are the final and initial state in the
       LHS and RHS of the scheme respectively.

States, *state–variables* and initial *state* are defined in the same way as in the
case of the scheme for TL operator $\bigcirc$.

3.3.2.4 Translation of Schemes involving TL Operator $\Diamond$

Schemes involving TL operator $\Diamond$ are classified using the same type of structure
as in the previous case for TL operator $\Box$. Generally speaking, the translation
task must guarantee that after the occurrence of a certain state, the formula
A given immediately after symbol $\Diamond$ will eventually hold. This has proved to
be difficult. As in previous cases, the structure defined in scheme 3.3 gives rise
to four different arrangements. The translation of some of them have not yet
been implemented as discussed in the following paragraphs.

1. The RHS of the implication reduces to a state–formula p'' representing a *state* in the *a–machine* domain $(\ldots \rightarrow \Diamond p'')$. In this case, the LHS can represent either a state with an associated transition or only a state in the *a–machine* domain.

   (a) If the LHS of the scheme defines the state and a transition (i.e. $p \wedge \tau \rightarrow \Diamond \ldots$), the translation procedure can be divided into two subtasks. Firstly, the *a–machine* domain is searched for a *state* q' equal to $h^{-1}(p'')$ or a (partially) covered *state* with the minimum number of covered *state–variables* covering $h^{-1}(p'')$. Secondly, a path is searched from the *state* $h^{-1}(p)$ to *state* q' where τ is the first output transition from $h^{-1}(p)$.

   If *state* q' is found, searching for a proper path can be done using standard path searching techniques (e.g. Williams (1990)). If, as a result of the search, it is concluded that such a path does not exist, generating one can be very difficult or even impossible. Here it is assumed that a path is always found. A proposed solution to overcome this problem, which has not been implemented, is to generate a Boolean quasi–lattice as described in the beginning of this subsection in which the glb is *state* $h^{-1}(p'')$ and the lub is *state* $h^{-1}(p)$. Again, the partial order assigned to the quasi–lattice is the partial covering of definition 2.6. In this way, it is guaranteed that all the possible behaviour between *state* $h^{-1}(p)$ and *state* $h^{-1}(p'')$ is considered and therefore the glb will be eventually reached.

   (b) If the LHS of the scheme defines only the state $h^{-1}(p)$ (i.e. $p \rightarrow \Diamond \ldots$), then the translation procedure is executed as in the previous case but

$$\delta(\sigma, h^{-1}(p)) \text{ is defined only for } \sigma \in \Sigma \text{ such that } |\sigma| = 1$$

   A special case occurs when the input transitions to *state* q' generate one or some of the desired *state–variable* values for *state* q''. In this particular *state* q' these *state–variables* are partially covered. If no transitions related to these *state–variables* appear in the self–loop of *state* q'' and no conflict with other input transitions exist, then these *state–variables* can be refined to their assigned values without introducing ambiguities.

2. For the scheme in which the RHS of the implication is modelled as a formula A, it is checked if the formula holds in the current *a–machine* structure. If the formula is true, then the initial state of the formula is

taken as p'' and the translation is performed as indicated in the two previous cases. If formula A does not hold in the current $a-machine$ structure, then a model that makes the formula to hold is generated independently using appropriate morphisms. Formula A is then connected to the structure corresponding to the LHS using the appropriate $states$ as in the case of the □ scheme.

$States$, $state-variables$ and initial $state$ are defined as in previous schemes. Once the translation is finished, the initial $state$ q_0 must be chosen from the generated structure as the most refined $state$ covering or being equal to the system initial state. The proposed mechanism for refining forbidden states and homomorphisms for translation into the $a-machine$ domain were implemented as PASCAL procedures. The implementation also handles negations in the "next transition" symbol when it appears in the RHS of any scheme. A negation of the "next transition" means that the transition must not be executed from the current state. This is achieved by erasing this transition in the current state. Examples are presented in the following section.

3.4 Modelling of Specifications for the Burner System

The burner system introduced in chapter 2 is used to illustrate the use of logic formalisms in the construction of formal models from given specifications. States to be avoided (forbidden states) during the operation of the burner and dynamic characteristics to be imposed on the system behaviour during the startup and shutdown procedures are described in an informal, descriptive way in tables 3.2 and 3.3. These informal specifications are assumed to be supplied by the user. This example is studied again in chapter 5 and extended to include emergency and abnormal operations.

S1.-	Avoid operating the igniter unless necessary.
S2.-	Avoid using fuel unless necessary.
S3.-	If the air valve is closed, the flame must never occur.

Table 3.2: Static specifications for the burner system.

In the system initial state, both valves are closed and the igniter is off. The startup operation is completed when both valves are open, flame has been detected and the igniter is off. The shutdown operation brings the system to its initial state leaving it prepared for restart.

D1.-	To start the system, first open the air valve, then switch igniter on, then open the fuel valve.
D2.-	Once air and fuel valves are open and the igniter is on, if flame is then detected, turn off the igniter. Otherwise close the fuel valve and switch the igniter off.
D3.-	After the flame is detected, the system will eventually be shut down. This must be done by closing the fuel valve and then the air valve.

Table 3.3: Dynamic specifications for the burner.

State–formulas describe the status of each of the four system components in the following order:

(air valve position, fuel valve position, igniter status, flame detector status)

When writing state–formulas, the numerical values introduced in table 2.1 are used to describe *state–variables*. In the text, the *state–variables* values will be referenced by their declarative description (i.e. open, closed, on, off).

3.4.1   Static Specifications

All specifications stated in tables 3.2 and 3.3 are vague. The translation into formal logic representations helps to overcome this problem. Specification S1 requires engineering interpretation of the "unless necessary" clause before it can be translated into a specific predicate statement. The safest procedure for starting up the burner is to minimise the time in which the air/fuel mixture is present in the chamber without the occurrence of flame. Therefore, if the air valve is closed, the igniter must under no circumstances be on, regardless of the status of other system components. This is expressed in a compact way by the following statement:

$$s_1 : (0, \infty_2, 1, \infty_4) = \text{ FALSE} \tag{3.4}$$

Note that the air valve and igniter values are always "closed" and "on" respectively while the fuel valve and the flame detector take all their possible combinations expressed by ∞_2 and ∞_4. The predicate in the LHS of this statement is mapped into the *a–machine* domain using isomorphism h and refined into the following *states*:

$$(0,0,1,0) \tag{3.5.a}$$
$$(0,0,1,1) \tag{3.5.b}$$
$$(0,1,1,0) \tag{3.5.c}$$
$$(0,1,1,1) \tag{3.5.d}$$

Upon examination, plausible interpretations can be assigned to each *state*. In *states* 3.5.a and 3.5.b, the igniter may be on when the air and fuel valves are closed with or without the presence of the flame, or when only the fuel valve is open with or without the flame being present as shown in *states* 3.5.c and 3.5.d. In each case the igniter operates when it is not required causing premature wear. Further inspection of the system model in fig. 2.6 shows that there is no other *state* in which the igniter operation is undesirable.

The meaning of specification S2 in table 3.2 is also ambiguous. A more precise formulation is that the fuel valve must be open only when the air valve is open because either the system is starting up and the igniter is on, complying with dynamic specification D1 of table 3.3, or the system is operating normally. This means that the fuel valve must never be open if the air valve is closed irrespective of the state of any of the other process components. This is described by statement s_{2a}:

$$s_{2a} : (0, 1, \infty_3, \infty_4) = \text{FALSE} \tag{3.6}$$

The following *states* result from the refinement of statement 3.6 and it is easy to see that all of them are undesirable:

$$(0,1,0,0) \tag{3.7.a}$$
$$(0,1,0,1) \tag{3.7.b}$$
$$(0,1,1,0) \tag{3.7.c}$$
$$(0,1,1,1) \tag{3.7.d}$$

A situation not considered by statement 3.6 is when the air and fuel valves are open without the presence of the flame and without the igniter being on. This situation is captured by the following statement:

$$s_{2b} : (1, 1, 0, 0) = \text{FALSE} \tag{3.8}$$

This statement is translated to the following *state*:

$$(1,1,0,0) \tag{3.9}$$

Specification S3 links the occurrence of the flame to the air valve being opened despite the status of the fuel valve and igniter. This specification is captured by the following statement in which, despite the state of the fuel valve or the igniter, it is prescribed that the air valve must not be closed if flame is present in the system:

$$s_3 : (0, \infty_2, \infty_3, 1) = \text{FALSE} \qquad (3.10)$$

The above statement covers the following states:

$$(0, 0, 0, 1) \qquad\qquad\qquad\qquad\qquad\qquad (3.11.\text{a})$$

$$(0, 0, 1, 1) \qquad\qquad\qquad\qquad\qquad\qquad (3.11.\text{b})$$

$$(0, 1, 0, 1) \qquad\qquad\qquad\qquad\qquad\qquad (3.11.\text{c})$$

$$(0, 1, 1, 1) \qquad\qquad\qquad\qquad\qquad\qquad (3.11.\text{d})$$

The four statements 3.4, 3.6, 3.8 and 3.10 modelling the forbidden states are shown in table 3.4. They cover in total 8 different forbidden *states* out of the 16 *states* that comprise the whole process model.

Forbidden States Statements		
Spec S1	s_1	$(0, \infty_2, 1, \infty_4)$
Spec S2	s_{2a}	$(0, 1, \infty_3, \infty_4)$
	s_{2b}	$(1, 1, 0, 0)$
Spec S3	s_3	$(0, \infty_2, \infty_3, 1)$

Table 3.4: Forbidden states arising from the static specifications.

3.4.2 Dynamic Specifications

Specification D1 of table 3.3 refers to sequential behaviour related to the process startup. Emphasis is given to the occurrence of events rather than to states. The initial system state is not explicitly declared in the specification statement. In this case, it will be inferred during the controller synthesis from the information provided. This is possible due to the simplicity of the example. In a more general case, as will be shown in section 5.4.2, the initial state of the system must be unambiguously specified. In order to use only the information given in the specification statement, all *state–variables* are assumed to be covered. Thus a TL formula can be written in which the sequencing of such events is specified using the TL symbol $\bigcirc$.

$$[(\infty_1, \infty_2, \infty_3, \infty_4) \quad \wedge \quad (\tau = 11)] \to [\bigcirc(\tau = 31)] \to [\bigcirc(\tau = 21)] \quad (3.12)$$

The meaning of formula 3.12 is as follows: from any state in the operation, if the air valve is opened ($\tau = 11$), the igniter must be switched on ($\tau = 31$) and the fuel valve must be opened ($\tau = 21$) in sequence immediately afterwards. No other transition can be executed before these two are executed.

The statements in the initial TL state are covering symbols for each elementary component of the system. The presence of the $\wedge$ symbol indicates that "next transition" $\tau = 11$ is not the only one that can occur from the covered *state*. The system behaviour can follow other trajectories. The formula only prescribes the behaviour following transition 11, that is, the immediate execution in sequence of transitions 31 and 21.

Having defined the initial TL state as covered, it is translated to a covered *state* in which its self–loop is labelled with all the executable transitions from elementary models with the exception of transition 11. If transition 11 is executed, the behaviour described by the formula must be enforced. Application of the scheme 3.1 to formula 3.12 results in the following subformulas:

$$(\infty_1, \infty_2, \infty_3, \infty_4) \wedge (\tau = 11) \to \bigcirc p' \qquad\qquad (3.13.\text{a})$$

$$p' \to \bigcirc(\tau = 31) \qquad\qquad (3.13.\text{b})$$

$$p'' \to \bigcirc(\tau = 21), \text{ such that } \delta(31, h^{-1}(p')) = h^{-1}(p'') \qquad (3.13.\text{c})$$

$$(3.13.\text{d})$$

The translation into the *a–machine* domain is performed by applying the appropriate mappings of subsection 3.3.2.2 obtaining the *a–machine* shown in fig. 3.2. The first state in the formula 3.12 describes a covered *state* in the *a–machine* domain. It can be observed in fig. 3.2 that the self–loop in *state* 1 contains all the possible system transitions except transition 11, making subformula 3.13.a true. Transition 11 is then connected to a *state* from which the sequence $\tau = 31$, $\tau = 21$ is executed satisfying subformulas 3.13.b and 3.13.c.

State 1 in fig 3.2 covers the initial/final and normal operation points. These were defined as marked *states* when constructing the process model in subsection 2.7.1. Therefore, they must be marked in the *a–machine* modelling this specification. The non connected input arrow in *state* 1 indicates that this is the initial state of the system.

Specification D2 describes the conditions in which the flame must occur. Once the air and fuel valves are open and the igniter is on, the flame may

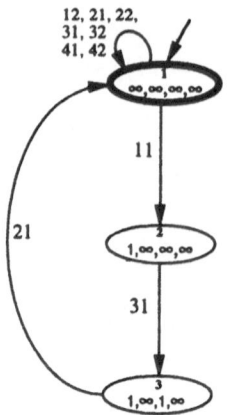

Figure 3.2: *a-machine* corresponding to specification D1.

be detected or not. Under normal circumstances ignition will occur and the flame will be detected, but abnormal situations must be considered as well. Flame might not be detected due to various faults which are beyond the present analysis. In case of the flame not being detected, the specification prescribes the execution of two elementary events, namely close the fuel valve and switch the igniter off. The specification does not prescribe any order. Here, an engineering decision must be introduced to resolve the conflict. In this case, transition 22 is executed before transition 32 to diminish the period of time in which the air/fuel mixture is present in the chamber. The following formula describes the required dynamic behaviour:

$$(1,1,1,0) \rightarrow$$
$$[\{[\bigcirc(\tau = 41)] \rightarrow [\bigcirc(\tau = 32)]\} \vee$$
$$\{[\bigcirc(\tau = 22)] \rightarrow [\bigcirc(\tau = 32)]\}] \tag{3.14}$$

This formula includes an exclusive disjunction prescribing the occurrence of two different trajectories from the same state. The formula is composed of two subformulas in which the TL operator $\bigcirc$ appears.

$$(1,1,1,0) \rightarrow [\bigcirc(\tau = 41)] \rightarrow [\bigcirc(\tau = 32)] \tag{3.15.a}$$
$$(1,1,1,0) \rightarrow [\bigcirc(\tau = 22)] \rightarrow [\bigcirc(\tau = 32)] \tag{3.15.b}$$

In this case, a refined state is specified at the beginning of TL formula

3.14. Therefore, this state is generated first in the *a–machine* domain using the procedure described in subsection 3.3.2.1. This gives rise to a quasi–lattice with 16 *states* where the lub is *state* $(\infty_1^1, \infty_2^1, \infty_3^1, \infty_4^0)$ and glb is the specified *state* $(1,1,1,0)$, the lub complement. From the glb *state*, the only prescribed transitions to be executed are transitions 41 and 22. This forces the elimination of all output transitions from the glb *state* $(1, 1, 1, 0)$ except these two. The final result is shown in figure 3.3.

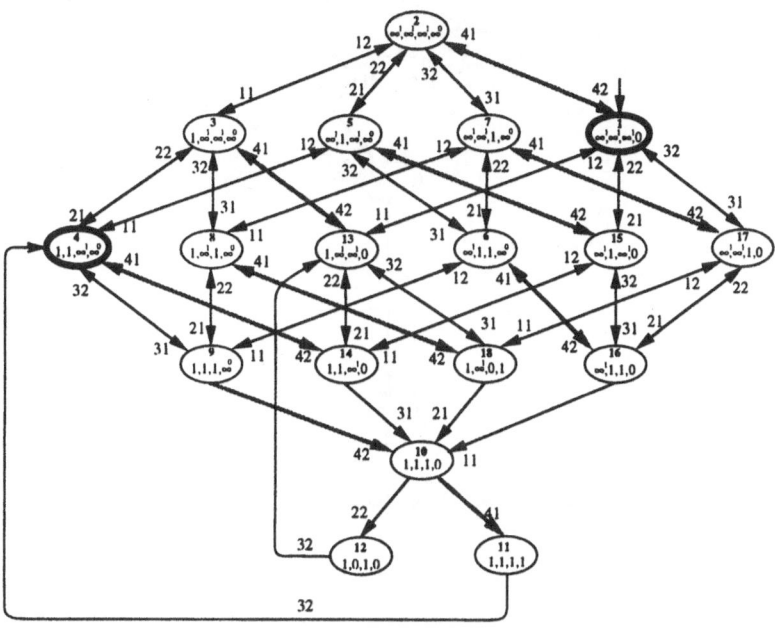

Figure 3.3: *a–machine* corresponding to specification D2.

Specification D3 describes the fundamental part of the shutdown process. Assuming that the system is operating normally, the shutdown of the system is effected by closing the fuel valve thereby extinguishing the flame. Finally, the air valve is closed. This is represented by the following TL formula:

$$(1, 1, 0, 1) \to [\bigcirc(\tau = 22)] \to [\bigcirc(\tau = 42)] \to [\bigcirc(\tau = 12)] \qquad (3.16)$$

For this TL formula, as in specification 2, an initial refined state is defined, $(1,1,0,1)$. Once it has been created as the glb of a quasi–lattice the rest of the lattice is constructed using the mapping for the TL operator $\bigcirc$. Note that a formula in which the eventual closing of the fuel valve is prescribed, $(1,1,0,1) \to \Diamond(1,0,0,1)$ would be true considering the quasi–lattice of the first part of TL formula 3.16 as the appropriate model. The final structure is shown in fig. 3.4.

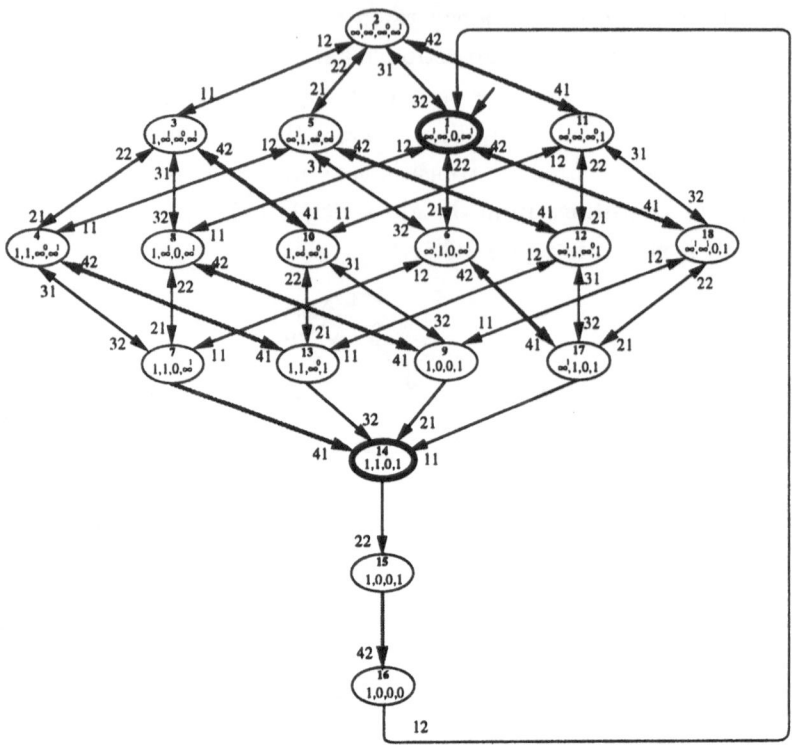

Figure 3.4: *a–machine* corresponding to specification D3.

3.5   Summary

The synthesis of supervisors and procedural controllers requires a model of the
process and a model of the desired behaviour to be imposed on the process.
Ways of dealing with static specifications have been proposed in the past (Ramadge and Wonham, 1987a; Kumar *et al.*, 1991a). However, no methods have
been developed for the construction of specification models considering dynamic behaviour. This has proved extremely difficult and error–prone even for
simple examples. In this chapter, logic formalisms are exploited together with
the modelling tools proposed in chapter 2 to efficiently construct behaviour
specifications that can be used in the controller synthesis.

Behaviour specifications are classified according to the logic formalisms used
for their modelling. Logic invariant specifications, in this case restricted only
to forbidden states, are modelled as Predicate Logic statements. Specifications
involving dynamic behaviour are modelled using linear Temporal Logic (TL)
formulas.

Forbidden states are translated into *a–machine states* using the method proposed by Ramadge and Wonham (1987a). The covering symbols for *state-variables* introduced in section 2.2 are used to facilitate this task. The treatment can be extended to other logic–invariant properties such as buffer capacities or finite counting.

In the case of TL formulas, these are constructed using a restricted syntax defined by three schemes for which homomorphisms are developed to map TL formulas into the *a–machine* domain. These homomorphisms give sufficient conditions to guarantee the existence of an *a–machine* for each TL formula which is valid in the behaviour set (i.e. TL frame) generated by the *asynchronous product* of the elementary components of the process being modelled. The underlying idea is to use the same elementary process components to construct an *a–machine* equivalent to the formula being translated. If the TL formula can be translated into the *a–machine* domain, it means that it is valid in the TL frame. Computational procedures were implemented in PASCAL to perform automatically this translation. The burner example introduced in chapter 2 was employed to illustrate the generation of specifications. Lists of forbidden states and dynamic specifications were given *a–priori* in natural language and in a rather imprecise manner. The refinement of this information to obtain the formal specifications is a task that must be performed by the engineer based on his own judgement. Once the Predicate Logic statements have been instantiated into propositional statements and the TL formulas have been obtained, the translation process into *a–machines* is done automatically. In order to illustrate the mechanisms of translation, only TL operator $\bigcirc$ was employed. Examples of the use of $\square$ and $\diamond$ operators will be shown in chapter 5 and 6.

In order to fully exploit the power of these logic formalisms in specification modelling, analysis tools need to be developed. For example, redundant or contradictory specifications must be identified in early stages of the synthesis process. Furthermore, establishing whether a system is completely specified is still an open issue.

Chapter 4

Supervisory Control Theory

4.1 Introduction

The overview of the Supervisory Control Theory given in chapter 1 was divided into three sections:

- Controllability and supervision.

- Control under partial observations.

- Decentralised and hierarchical control.

In this chapter, the controllability and supervision section is revisited in detail. It presents the fundamental control concepts for DESs from which modifications are proposed in order to enable the synthesis of procedural controllers for process systems. As mentioned in the Preface, *supervision* in Supervisory Control Theory is understood as the action of maintaining the closed–loop behaviour of a given process within certain boundaries by disabling the execution of manipulable (i.e. controllable) events. However, the execution of an operating procedure in a process plant involves not only the disabling of certain events but actually the enforcement of others (i.e. to exert control actions upon the system). In this respect, Supervisory Control Theory lays the foundations for characterising the closed–loop behaviour generated by the supervisor and process as the set of all possible controllable behaviours, but no mechanisms are given to enforce a particular behaviour. Generally speaking, that is the role of a controller. The synthesis of controllers will be the subject of the next chapter.

The Supervisory Control Theory has been evolving since the first concepts were proposed in the PhD thesis of Peter Ramadge in 1983. This work uses concepts and nomenclature as presented in Wonham (1988). First, section 4.2 introduces the original concepts of *supervisor* and *control action* proposed in Supervisory Control Theory. Then, in the following section, definitions of different types of supervisors are presented. Of particular interest is the *proper* supervisor which forces the closed–loop behaviour to avoid blocking conditions. The concept of *controllability* is also introduced and sufficient conditions for the existence of proper supervisors are given. After discussing the existence of supervisors, two methods for calculating controllable behaviours (i.e. languages) originally proposed by Wonham and Ramadge (1987) are discussed. Having revised the basis of Supervisory Control Theory, section 4.4 shows why the given definition of controllability results in conservative conditions for the existence of controllable behaviour in the case of chemical processes. After this, a novel definition of a relaxed controllability, *conditional controllability*, is introduced together with the matching definition of a supervisor. Sufficient conditions for the existence of this supervisor are established. A method for the calculation of languages satisfying conditional controllability is provided in the same section. Issues concerning supervisor modularity proposed in the basic Supervisory Control Theory are explored to ease the synthesis procedure proposed in the following chapter. Finally, the chapter closes with a summary.

4.2   The Feedback Supervisory Model of Supervisory Control Theory

In Supervisory Control Theory, the process is represented by a *FSM P* called the generator (Ramadge and Wonham, 1987b):

$$P = \{Q, \Sigma, \delta, q_0, Q_m\}$$

where

Q Set of states.

Σ Set of transitions, which is divided into controllable (Σ_c) and uncontrollable (Σ_u) transitions.

δ Transition function defined as a partial function $\delta : \Sigma \times Q \to Q$.

q_0 Initial state.

Q_m Set of marked states.

The feedback supervisor S of Supervisory Control Theory is composed of an *FSM* S_{sur} and a mechanism Ψ identified as the state feedback map:

$$S = (S_{sur}, \Psi)$$

The FSM S_{sur} is defined as the 5-tuple

$$S_{sur} = \{X, \Sigma, \xi, x_0, X_m\}$$

where

X Set of states.

Σ Set of transitions.

ξ Transition function defined as a partial function $\xi : \Sigma \times X \to X$.

x_0 Initial state.

X_m Set of marked states.

and the state feedback map Ψ which is a total function given by

$$\Psi(\sigma, x) = \begin{cases} 0 \text{ or } 1 & \text{if } \sigma \in \Sigma_c \\ 1 & \text{if } \sigma \in \Sigma_u \end{cases}$$

The function Ψ is implemented together with a mechanism to enable/disable the occurrence of controllable transitions in the process. If $\Psi(\sigma, x) = 0$ and $\sigma \in \Sigma_c$, the transition σ in the process is disabled. Otherwise, it is enabled. When $\sigma \in \Sigma_u$, the function Ψ is always equal to 1 and thus σ is enabled.

In the original Supervisory Control Theory, P and S are synchronised on the set of transitions Σ shared by both. The states on S are mapped using the state feedback Ψ into "control patterns" as indicated above, enabling or disabling the execution of controllable transitions in the process P (see fig. 4.1). The supervisor does not generate control "commands". Instead, it disables controllable transitions that are not allowed to occur. The final decision of what enabled transition to execute depends entirely on P.

The combined action of process and supervisor generates a subset of process trajectories (a sublanguage) identified as the closed–loop language $L(S/P)$. A

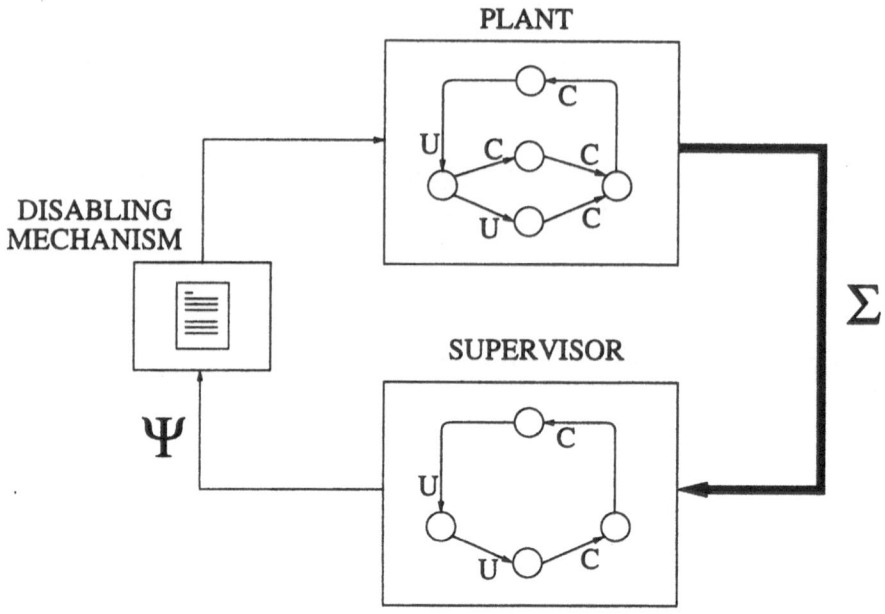

Figure 4.1: Feedback control mode arrangement in Supervisory Control Theory.

marked language $L_m(S/P)$ is also associated with it (Ramadge and Wonham, 1987b):

$$L(S/P) = L(S) \cap L(P) \text{ and}$$
$$L_m(S/P) = L_m(S) \cap L_m(P)$$

The closed–loop language describes the process overall behaviour while the marked language only considers closed–loop trajectories ending in one of the previously marked *states*. If the marked *states* are initial or final points of operation, safe states or conditions in which the process can be halted, then the closed–loop marked language will contain trajectories that the process must achieve by interacting with the controller.

The main advantage of this representation is its generality. The model is sufficiently abstract to permit the derivation of more complex constructs from it as required. However, the imposed full synchronisation between the process and the supervisor may not be commonly found in practice. It is more likely that a set of key process variables be measured, from which a control action is decided.

4.3 Controllability and Supervisors in Supervisory Control Theory

The original supervisor definition accounts, in a very general fashion, for the use of the feedback function as the means of realizing supervisory actions. A supervisor S for a process P that can be represented by its state–transition structure alone (i.e. without explicit mention of the control function Ψ) is called a *neat* supervisor. In other words, the transition function only exists for transition σ and state x such that $\Psi(\sigma, x) = 1$. It is also convenient to guarantee that the supervisor will always consider any trajectory belonging to the closed–loop system. In other words, that the supervisor is *closed* with respect to the closed–loop language. Assume that $L(S/P)$ is the closed–loop language. The supervisor, in order to be *complete* must be designed so that if a given trajectory exists in the closed–loop language, then the occurrence of an uncontrollable transition is still part of the closed–loop language (Ramadge and Wonham, 1987b). Furthermore, it is also desirable to have a trim structure as a supervisor. A supervisor that is trim, neat and P–complete is called a *standard* supervisor. Finally, it may be the case that the supervisor blocks marked *states* in the closed–loop language. A supervisor that avoids this problem is a *proper* supervisor:

Proposition 4.1 *Proper Supervisor*
  A standard supervisor S is proper if and only if S/P is nonblocking:

$$\overline{L_m}(S/P) = L(S/P)$$

i.e. every trajectory of the closed loop process can be extended to reach the set of marked states.

In this work any supervisor will be considered *proper* unless stated otherwise. As was shown in chapter 2, in order to be sure that all marked *states* can be reached in $L(S/P)$, $L_m(S)$ and $L_m(P)$ must be nonconflicting:

$$\overline{L_m}(S/P) = \overline{L_m(S) \cap L_m(P)} = \overline{L_m}(S) \cap \overline{L_m}(P) \qquad (4.1)$$

i.e. $L_m(S)$ and $L_m(P)$ are nonconflicting languages and S/P is trim (because all states must be reachable and coreachable).

This is formalised in the following theorem.

Theorem 4.1 *If $L_m(S)$ and $L_m(P)$ are nonconflicting, then S/P is nonblocking, that is, $L(S/P) = \overline{L}_m(S/P)$ (Wonham, 1988).*

Proof

Assume

- $L(S)$ and $L(P)$ are closed.

- S and P are trim.

If $L_m(S)$ and $L_m(P)$ are nonconflicting then

$$
\begin{aligned}
\overline{L}_m(S/P) &= \overline{L_m(S) \cap L_m(P)} & \text{by definition}\\
&= \overline{L}_m(S) \cap \overline{L}_m(P) & \text{nonconflicting property}\\
&= L(S) \cap L(P) & \text{trim definition}\\
&= L(S/P)
\end{aligned}
$$

Thus $\overline{L}_m(S/P) = L(S/P)$, and $L(S/P)$ is nonblocking. q.e.d.

4.3.1   Controllability and Existence of Supervisors

In differential calculus–based systems, it is said that a system is controllable if it is possible to find a "control action" such that the origin state of the system can be reached from any initial state in a finite time. Related to controllability is the concept of stability. A system is stable if it has a bounded response. That is, if the system is subject to a bounded input or disturbance and the response is bounded in magnitude, then the system is stable.

For state–event based systems, the idea of reaching an origin is seldom encountered and if so, it is not defined in terms of continuous–time variables with numerical values. In the Supervisory Control Theory the concept of controllability considers the possibility of maintaining the system within a well defined behaviour space:

Definition 4.1 *Controllability.*

A nonempty language $K \subseteq L_m(P) \subseteq \Sigma^*$ is said to be controllable with respect to a machine P if

$$
\overline{K}\Sigma_u \cap L(P) \subseteq \overline{K}
$$

In other words, a language is controllable if the uncontrollable transitions never cause a string to exit the prefix closure $\overline{K}$ (Ramadge and Wonham, 1987b).

The existence of a proper supervisor S which, together with a process P, generates a language K that is controllable, was first proved by Ramadge and

Wonham (1987b). The nonconflicting property between $L_m(S)$ and $L_m(P)$ was later included to guarantee nonblocking (Wonham, 1988)[1].

Theorem 4.2 (Existence of a proper supervisor) *(Theorem 6.1 in Ramadge and Wonham (1987b) and theorem 4.1 in Wonham (1988))*
If $K \subseteq L_m(P)$ is a controllable language and $L_m(P)$ and $L_m(S)$ are non-conflicting then S is a proper supervisor such that $L_m(S/P) = K$.

4.3.2   Computation of Controllable Languages

Ramadge and Wonham (1987b) define the union of all controllable sublanguages of a given language L as the *supremal controllable sublanguage* $K^\uparrow$

$$K^\uparrow := \cup\{K : K \subset L \text{ and } K \text{ is controllable}\}$$

They proved that $K^\uparrow$ is well defined and unique. That is, $K^\uparrow$ is closed under arbitrary unions and therefore it contains all the controllable sublanguages on a given language L. Moreover, if K and L are regular, then $K^\uparrow$ is also regular. The supervisor that generates this language is identified as the *minimal restrictive* supervisor.

Wonham and Ramadge (1987) proposed two methods for the computation of regular controllable languages and their associated *FSMs*. The first one characterises the supremal controllable language $K^\uparrow$ as the largest fixed point of the function $\Omega : \Gamma \to \Gamma$ defined as

$$\Omega(K) = L \cap sup\{T : T \subset \Sigma^*, T = \overline{T} \text{ and } T\Sigma_u \cap L' \subseteq \overline{K}\}$$

where $K \in \Gamma$
Γ is the power set of Σ^*
L and L' are fixed languages
T is a language which is closed and controllable on K.

If L is regular, the supervisor S can be calculated as the limit of a (finite) sequence of languages K_j given by

$$K_{j+1} = \Omega(K_j), \; j \geq 0, \; K_0 = L$$

The second approach, which is used here, considers only closed languages modelled by *FSMs*. This is also an iterative calculation process in which,

---

[1] Proof of existence for complete supervisors is given in Ramadge and Wonham (1987b, theorem 5.1). A simpler proof considering full synchronisation between S and P is given in Kumar *et al.* (1991a, Lemma 2.7).

given an approximation of the controllable language and its associated *FSM*, states from where the language becomes uncontrollable are eliminated. This is explained in detail in the following paragraphs.

First, assume the existence of $P = \{Q, \Sigma, \delta, q_0, Q_m\}$ and $C_j = \{X, \Sigma, \xi, x_0, X_m\}$ the *FSMs* that realise the process language and the approximation K_j to the controllable language respectively. Then, define the following:

- A unique mapping $f : X \to Q$, where f indicates the correspondence between the states of the process P and the approximation C_j.

- The *active set of state* q, $\Sigma(q) \subset \Sigma$, as a subset of transitions for each state q in which $\delta(\sigma, q)$ is defined.

- The *activity event constraint*, $(\Sigma[f(x)] \cap \Sigma_u) \subset \Sigma(x)$, for each *state* x on C_j.

The *activity event constraint* is checked for each *state* in the approximation C_j. If a *state* x on C_j does not satisfy it, it means that there exists a *state* in the process P which may execute an uncontrollable transition not considered in C_j and therefore that string fails to be controllable. Consequently, the construction of a *FSM* which realises the maximal controllable language can be done by removing those *states* in C_j that fail to satisfy the *activity event constraint*. This *FSM* realises the supremal controllable language $K^\uparrow$ (Wonham and Ramadge, 1987). From here on, when mentioning any controllable language it will be assumed that it is $K^\uparrow$ unless indicated otherwise.

Kumar *et al.* (1991a) presented an alternative definition of controllability for closed languages and it is used to propose a different algorithm for the calculation of $K^\uparrow$. These two last algorithms are based on the same principles and apparently there is no particular advantage of one over the other.

4.4 Conditional Controllability and Supervisor Synthesis

Summarising the previous sections, in the Supervisory Control Theory a nonempty language $K \subseteq L_m(P) \subseteq \Sigma^*$ is said to be controllable with respect to an *a–machine* P if $\overline{K}\Sigma_u \cap L(P) \subseteq \overline{K}$, that is, if an uncontrollable event never causes a string to exit the prefix closure $\overline{K}$. If $K \subseteq L_m(P)$ is controllable and $L_m(P)$ and $L_m(S)$ are nonconflicting then S is proper and $L_m(S/P) = K$. As was shown in section 4.3.2, two methods for the calculation of the maximum controllable language and its associated *FSM* that follows from the controllability definition were introduced in Wonham and Ramadge (1987).

When dealing with systems where transitions model uncontrollable actions that can occur unexpectedly, such as sensor signals, synthesising a supervisor for a given specification as stated in Supervisory Control Theory can be a difficult task. The controllability definition enforces conservative conditions by assuming that any existing uncontrollable transition in a given *state* will be executed and, therefore if it leads to uncontrollable behaviour, this *state* must be eliminated from the controllable space. Thus, a suitable supervisor must account for all the occurrences of these uncontrollable transitions with the subsequent appropriate behaviour guaranteeing controllability. This implies that, somehow, the corresponding legal behaviour must consider all these circumstances in advance. Although desirable, the design of a supervisor of this nature in a one–step approach is almost impossible to achieve for any system of realistic size and complexity. For instance, in the burner example whose process model and static and dynamic specifications were developed in chapter 2 and 3 respectively, the flame can be detected at any moment as part of either the normal operation or abnormal behaviour. A supervisor considering only normal operation would contain the transition that detects flame only after the appropriate circumstances occur during the startup (air and fuel valves open and igniter on). The supervisor would be blind to flame detection under any other circumstance as it is shown in the examples of chapter 5. The inherent indeterminacy of the *FSMs* imposes the need for their analysis making the supervisor design an iterative process.

It has been found, as will be shown in chapters 5 and 6, that the calculations using the concept of controllability leads very frequently to maximum controllable languages that are of no practical interest or even empty. Therefore this definition is relaxed to characterise a bigger language that covers the systems where controllability is subject to the conditional execution of controllable transitions. The idea is to consider the possibility of having supervisor states in which controllable transitions are executed while in the corresponding process state $f(x)$, uncontrollable transitions may exist.

Let $\Sigma_{con} \subseteq \Sigma_u$ be a subset of uncontrollable transitions that may exist in any state. These uncontrollable transitions, that should be identified during the construction of the process model, allow the conditional existence of a controllable language as shown in the following paragraphs.

Definition 4.2 *Conditional Controllability.*

A language $K_{cc} \subseteq L_m(P) \subseteq \Sigma^*$ is said to be conditionally controllable with respect to a FSM P if $\overline{K}_{cc}\sigma_u \cap L(P) \subseteq \overline{K}_{cc}\sigma_{con}$, and $\sigma_u \in \Sigma_u$, $\sigma_{con} \in \Sigma_{con}$.

In other words, a given language is *conditionally controllable* if an uncontrollable transition never causes a string to exit the language prefix closure followed by any of the uncontrollable transitions from Σ_{con}. This relaxation of the definition of controllability leads to the definition of a proper supervisor S_c, in which the marked language is the conditional controllable language of interest.

Theorem 4.3 *Existence of a Conditional Supervisor.*

A proper supervisor S_c for a process P exists such that $L_m(S_c/G) = K_{cc}$ if K_{cc} is conditionally controllable and $L_m(P)$ and $L_m(S_c)$ are nonconflicting. Such a supervisor is called a conditional (proper) supervisor.

Proof
Assume

- S_c and P are trim.

- $L(S_c)$ and $L(P)$ are closed.

- $K_{cc} = L_m(S/P)$.

If K_{cc} is conditionally controllable then

$$
\begin{aligned}
\overline{L}_m(S_c/P)\Sigma_u \cap L(P) &\subseteq \overline{L}_m(S_c/P)\Sigma_{con} && \text{by definition} \\
\overline{L}_m(S_c/P)\Sigma_{con} &= [\overline{L_m(S_c) \cap L_m(P)}]\Sigma_{con} && \text{by definition} \\
&= [\overline{L}_m(S_c) \cap \overline{L}_m(P)]\Sigma_{con} && \text{nonconflicting} \\
&= [L(S_c) \cap L(P)]\Sigma_{con} && \text{trim} \\
&= L(S_c/P)\Sigma_{con} && \text{by definition. q.e.d.}
\end{aligned}
$$

K_{cc} is computed by relaxing the method used here for calculating a controllable language. Assume P and C_j are the *FSMs* corresponding to the process and the approximation to the conditionally controllable language. If a *state* $x \in X$ does not meet the *activity event constraint* it means that there are uncontrollable transitions in the *state* of P that are not present in the active set $\Sigma(x)$ of the corresponding *state* of C_j representing the supervisor under synthesis.

If $\Sigma(x) - [\Sigma(x) \cap \Sigma_u] \neq \emptyset$ in the *state* $x \in X$ where the activity event constraint failed, then there exist controllable transitions in the *state* that can lead the system to a conditionally controllable language provided that $(\Sigma[f(x)] \cap \Sigma_u) \in \Sigma_{con}$. Therefore, the construction of the *FSM* representing the conditional supervisor S_c is done by removing the *states* that fail to meet the activity constraint if

1. There are no controllable transitions in $\Sigma(x)$, or

2. The uncontrollable transitions in the current *state* x not belonging to the active set are not in the Σ_{con} set.

4.5 Modular Synthesis of Supervisors

When having multiple specifications for a given system, under certain circumstances, the supervisor synthesis task can be divided into several subtasks, each task synthesising a supervisor for a separate specification. Conditions for the existence of modular solutions were given in Wonham and Ramadge (1988). The modular synthesis simplifies the overall task of finding a global supervisor as well as modifying and maintaining it. It also reduces the computational complexity, especially when P is large.

The global supervisor is obtained using a conjunction operation on the individual supervisors called *supervisor conjunction*. The *supervisor conjunction* of two supervisors S and T is defined as the synchronous product of the two corresponding *a–machines* that realises the language intersection of their marked languages. Wonham and Ramadge (1988) proved that the family of complete supervisors for a given process is closed under *supervisor conjunction* and the conjunction operation is associative and commutative. A formal definition of this operation is given in the following proposition. The proof can be found in Wonham and Ramadge (1988).

Proposition 4.2 *Supervisor Conjunction* ($\sqcap$) *(Wonham and Ramadge, 1988, proposition 4.1).*

   S and T are complete supervisors for a given supervised process P. Then
1) $L(S \sqcap T/P) = L(S/P) \cap L(T/P)$
2) $L_m(S \sqcap T/P) = L_m(S/P) \cap L_m(T/P)$
3) $S \sqcap T$ is a complete supervisor for P.

It is important to note that the nonblocking property is not closed under this operation. Therefore, in order to guarantee that the conjunction of supervisors is nonblocking, Wonham and Ramadge (1988, proposition 4.2) proved that if supervisors S and T are nonblocking for a process P and $L_m(S/P)$ and

$L_m(T/P)$ are nonconflicting then it will always be the case that $S \sqcap T$ is non-blocking. Furthermore, it was observed that controllability is not closed under language intersection (or synchronous composition of the corresponding *FSMs*). Conditions to guarantee controllability are given in the following proposition.

Proposition 4.3 *Controllability of Language Intersection (Wonham and Ramadge, 1988, proposition 5.1)*

If $L_1, L_2 \subseteq \Sigma^*$ and if L_1 and L_2 are nonconflicting and controllable, then $L_1 \cap L_2$ is controllable.

Note that these are only sufficient conditions. L_1 and L_2 may not be controllable but $L_1 \cap L_2$ is controllable. L_1 and L_2 may be conflicting but $L_1 \cap L_2$ is controllable. Also the supervisor S that realises $L_1 \cap L_2$ is not necessarily nonblocking. Nevertheless, these propositions simplify the supervisor synthesis.

In summary, to find the nonblocking conjunction of two trim supervisors, it must be guaranteed that their corresponding closed–loop marked languages are nonconflicting.

4.6 Summary

Controllability and supervision issues of the Supervisory Control Theory have been reviewed. In particular, the following aspects were discussed:

- Control arrangement and role of the supervisor.

- Definition of a proper supervisor.

- Definition of controllability.

- Proof of existence for proper supervisors using sufficient conditions.

- Methods for calculating controllable behaviour.

- Conditions for modular supervisor synthesis, reducing computational complexity.

However, the concept of controllability makes the synthesis of supervisors for process systems difficult and even rules out behaviour of interest. To overcome this hurdle, the following additions were made:

- Proposal of the concept of *conditional controllability* to allow the existence of uncontrollable transitions from a given state that may not necessarily be executed.

- Definition of supervisors that meet conditional controllability and development of sufficient conditions for their existence.

- Development of a calculation method for conditional controllable behaviour extending the method proposed by Wonham and Ramadge (1987) for closed languages.

The language generated by the supervisor can be considered as a superset of controllable trajectories. In addition, the concept of conditional controllability permits the construction of supervisors realising larger languages. As will be shown in the next chapter, this is very useful for the synthesis of suitable controllers based on this supervisory structure.

To summarise, when synthesising supervisors one must ensure

- The marked languages of process and supervisor are not conflicting. This guarantees that the supervisor does not block marked states of interest in the process.

- to closed–loop marked languages from different supervisors must be non conflicting. This guarantees that trim supervisors do not block each other.

Computer procedures were implemented to perform all the required operations.

Chapter 5

Synthesis of Procedural Controllers

5.1 Introduction

In chapters 2 and 3, a consistent modelling framework was developed to construct process models and behaviour specifications. In chapter 4, modifications were introduced to Supervisory Control Theory to expand the class of systems amenable to supervision. However, Supervisory Control Theory was conceived to perform only supervisory activities an not to enforce control actions on a process.

In this chapter, a reinterpretation of the control arrangement of Supervisory Control Theory is proposed in which the synthesised *a–machine*, now a *local controller*, is able to execute commands in feedback mode in order to satisfy the behaviour prescribed by a dynamic specification. A *global controller* is then defined as the *a–machine* obtained from the *controller conjunction* (in the sense of Wonham and Ramadge (1988)) of *local controllers* satisfying the set of all dynamic specifications in which at most one control command is permitted to occur from each controller state.

In section 5.2, the proposed arrangement of the control system is introduced together with definitions of *local* and *global* controllers. The controllers and process are assumed to be synchronised by the transition set under an input–output interpretation. The *global* controller issues control commands to the plant as a function of previous controller states and executed transition. Section 5.3 presents the synthesis procedure to obtain such controllers.

First, the structure that characterises the (conditionally) controllable system behaviour that avoids the forbidden states is synthesised. This is equivalent to the supervisor proposed in Supervisory Control Theory which only disables controllable transitions. Using this structure as a basis, a global controller is constructed by a specification refinement process in which at most one controllable transition is permitted to occur from each controller state. When dealing with multiple dynamic specifications, the modular approach proposed by Wonham and Ramadge (1988) presented in subsection 4.5 is used to ease the synthesis procedure. Finally, in section 5.4, controllers are synthesised for two examples. In the first example, the objective is to obtain a global controller for startup, normal operation and shutdown of the burner system introduced previously in subsection 2.7.1. The synthesis of this small controller is graphically demonstrated. The use of Supervisory Control Theory yields a supervisory structure whose closed–loop behaviour (language) is empty. Using conditional controllability a suitable supervisor is then synthesised from which the controller is constructed. It also permits the identification of conditions in which the system can drift into abnormal or emergency situations. The second example expands on the result of the first to obtain a more comprehensive controller that considers normal, abnormal and emergency operations.

5.2   Control Arrangement

In a standard interpretation of process control, a system can have three types of variables: inputs, states and outputs. The input variables are manipulated or disturbance variables that enforce changes upon the system. The states describe how the system evolves. The outputs are used to monitor the status of the system. The outputs, and sometimes the states, particularly in chemical engineering literature, are associated with measured variables. This classification allows to identify the role of each process component in the system behaviour.

As mentioned in chapter 2, within the *a–machine* framework each *state–variable* is associated with an elementary component *a–machine*. The transition function γ associated with the elementary *a–machine* describes how this *state–variable* evolves. Within a standard control framework these *state–variables* can be interpreted as inputs, states or outputs according to their role in the system.

Transitions associated with input *state–variables* can be identified as input transitions. They describe how external actions are exerted upon the system. If they are controllable transitions, they can be interpreted as the "control commands" to manipulate the system. Any other input transitions is identified as input disturbance and thus uncontrollable. In a similar fashion, all the transitions associated with the model states and outputs are also uncontrollable.

In order to use an input–output model, mappings must be defined between transitions and the required inputs and outputs. The set of controllable transitions, Σ_c, corresponds to the input transitions executed on the plant hardware. Presently, disturbances are not considered. The set of uncontrollable transitions, Σ_u, are output transitions matched to the set of measurements taken from the plant using dedicated measuring devices. The controller is arranged in feedback mode as shown in fig. 5.1. The controller receives output transitions from the plant via the measurement instrumentation and executes them synchronously. Based on the measurements in each controller state, the controller issues one control command at most or executes the next output transition. It is assumed that measurements are available for all output transitions. In the rest of this and the next chapter, transitions will be identified using Supervisory Control Theory nomenclature. Controllable transitions will refer to control commands while uncontrollable transitions will be associated with measurements.

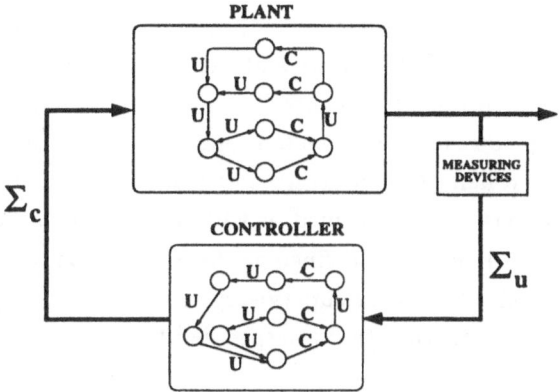

Figure 5.1: Feedback control mode arrangement.

5.2.1   Controller Definitions

Supervisory Control Theory provides a control mechanism to maintain the closed–loop process behaviour within a finite set of trajectories defined as controllable. This control mechanism is suitable for the compliance of specifications that require the avoidance of certain behaviour such as forbidden states. Dynamic specifications, on the other hand, prescribe the execution of specific events. The enforcement of such a behaviour in a process requires a mechanism capable of determining which control action to execute. A *local controller* is understood to be a structure describing the enforcement of certain behaviour prescribed by a dynamic specification:

Definition 5.1 *Local Controller.*

   A local controller is an a–machine realising the following behaviour (language) for a given specification SPEC:

$$L(C) = \quad sup \ \{ \quad T : T \subseteq (L(\text{SPEC}) \cap L(S)) \subset \Sigma^*,$$
$$T = \overline{T} \ and \ T \ is \ (conditionally) \ controllable \}$$

Thus the synthesis of either a local controller or a supervisor becomes the synthesis of a conditionally controllable or controllable language as defined in chapter 4. Therefore, in terms of language synthesis, there is no distinction between "supervisor" or "controller". In terms of interpretation, a "controller" realises languages defined by dynamic specifications, while a supervisor does the same for forbidden states. It was mentioned in section 4.3 that in order to guarantee the existence of a proper supervisor such that $L_m(S/P)$ is controllable, both $L_m(P)$ and $L_m(S)$ must be nonconflicting. Languages guaranteed to be nonconflicting were mentioned in section 2.5. These include prefix–closed languages or languages whose respective alphabets are disjoint. In general, this property must be ascertained. A controller can be obtained in the same fashion. Similarly, the *supervisor conjunction* operator can be used to obtain the conjunction of controllers. Moreover, from now on when a conjunction of either controllers or supervisors is performed, it will be referred as their conjunction.

A local controller language is defined as the intersection of the languages realised by the controller C and the supervisory structure S in which only strings from the supervisory structure that are specified by the controller survive such that

$$L(C) \ \ = L(C) \cap L(S) \ \ \ \ = L(C/S) \quad and$$
$$L_m(C) \ = L_m(C) \cap L_m(S) \ = L_m(C/S)$$

In this way it is guaranteed that the local controller will generate a closed–loop language that not only satisfies the given dynamic specification but also avoids all the forbidden states.

A local controller is guaranteed to be nonblocking, and therefore *proper* in the sense of proposition 4.1, if $\overline{L}_m(C)$ and $\overline{L}_m(S)$ are nonconflicting. The supervisory structure was defined as trim ($L(S) = \overline{L}_m(S)$) and thus, prefix–closed ($L(S) = \overline{L}(S)$). Thus, if the controller is defined to have the same properties, the existence of a *proper* controller can be guaranteed (see theorem 4.1).

Having obtained a *local controller* for each specification, a *global controller* is defined as the conjunction, in the sense of proposition 4.2, of all the *local controllers*, mapping each system state into at most one controllable transition. According to proposition 4.2, in order to guarantee the existence of a *global controller*, the marked languages of all *local controllers* in the conjunction must be nonconflicting. If a *global controller* contains *states* in which two or more controllable transitions exist, extra dynamic specifications must be included in order to eliminate all but one. This is done by generating local controllers for the new dynamic specifications and doing the *controller conjunction* with the existing global controller. At present no method is proposed to carry out this refinement process, leaving it entirely up to the user judgement.

5.3   Synthesis Procedure

The synthesis procedure requires the following input information:

- Elementary process components and their operational description. The elementary components can be obtained from a process diagram. The operational description must be given in terms of obtainable states and executable transitions of each elementary component.

- A list of specifications describing the desirable and undesirable system behaviour.

The procedural controller is synthesised in four steps:

- **Step 1.- Process Modelling**

  The process model P on which to base the controller is built from elementary process components that can be modelled individually as *a-machines*, as shown in chapter 2. Each elementary model must represent a process component in which process inputs or outputs transitions or both are executed. In each elementary model, transitions corresponding to control commands must be labelled as controllable while those corresponding to system responses are labelled as uncontrollable. Using these elementary models, the model for the complete process can be generated in various ways as shown in chapter 2. For instance, it can be obtained from the *asynchronous product* of all elementary process components. Also, causal relationships modelled by TL formulas which are translated into *a–machines* can be incorporated into the model using the *synchronous product*. In any given model, *states* of particular relevance to the operation must be marked. These *states* can be initial and final conditions of startup and shutdown operations, states in which the process can be halted, normal operation states or safe states. The final *a–machine* must be checked to be trim as described in definition 2.3.

- **Step 2.- Specification Modelling**

  Specifications are given, in the first instance, in natural language. Then, they are manually translated into logic representations to avoid ambiguities in the interpretation. As discussed in chapter 3 they are divided into two types:

  - Specifications concerning forbidden states that can be translated into PL statements.

  - Specifications related to dynamic characteristics which can be represented as TL formulas. These are identified as dynamic specifications.

  The translation of forbidden states and dynamic specifications (PL statements and TL formulas respectively) into the *a–machine* domain was presented in detail in chapter 3. Again, all the obtained *a–machines* must be checked to be trim. Also, relevant states for the system operation must be marked and must match with those in the process model. In this way the maximum number of marked *states* is guaranteed to exist in the supervisors and controllers obtained.

- **Step 3.- Supervisory Structure Synthesis**

  Once in possession of a process model P and the set of formal specifications describing all the behavioural characteristics to be imposed on the process, the set of uncontrollable specifications, $\Sigma_{con} \subseteq \Sigma_u$, originating uncontrollable behaviour must be identified. Then the supervisory structure S is synthesised by eliminating from the process model P the forbidden states defined by the static specifications and then obtaining the surviving (conditionally) controllable structure using the tools described in chapter 4. The supervisor languages are defined as

$$L(S) \;\;= L(S) \cap L(P) \;\;\;\;= L(S/P) \;\;\;\; \text{and}$$
$$L_m(S) \;= L_m(S) \cap L_m(P) = L_m(S/P)$$

The obtained supervisory structure is guaranteed to be proper (i.e. $\overline{L_m}(S) = L(S) = L(S/P)$) if S and P are trim. Again this property must be checked by the user.

- **Step 4.- Controller Synthesis**

  Dynamic specifications prescribe behaviour that must be enforced upon the system. First, a local controller is synthesised for each specification. Afterwards, the global controller is obtained as the conjunction of the individual controllers. As stated above, the resultant structure must map each *state* into one controllable transition at most. If the resultant *a–machine* contains *states* from which more than one controllable transition occurs, extra dynamic specifications must be added to eliminate the indeterminacy.

5.4   Examples

Two examples are presented as an introductory illustration of the synthesis procedure. The first example, the burner system introduced in subsection 2.7.1, serves to illustrate most of the ideas and concepts presented so far. Its small size allows one to follow most of the synthesis steps pictorially. The generated global controller is represented graphically and its compliance with the stated behaviour specifications can be checked visually. Conditional controllability, apart from being used for the normal startup and shutdown global controller synthesis, allows the identification of *states* from where the process may drift into abnormal, and very likely, uncontrollable situations. This poses the need to consider these *states* not as forbidden as in the initial specification but as starting points of procedures to handle abnormal behaviour that should be part of the behaviour specifications. The second example deals with this situation by including two alarm devices to indicate the undesirable presence or absence of flame in the system. The set of specifications is modified and augmented to consider emergency operation under abnormal behaviour. A global controller considering startup, shutdown, normal and abnormal operation is then synthesised.

5.4.1   Burner System

The first example is taken from Moon *et al.* (1992). The process is the simplified burner system introduced in subsection 2.7.1 and shown there in fig. 2.2. It consists of two on/off valves on the air and fuel feeds to the burning chamber, a flame igniter and a flame detector in the chamber. In the original reference, the objective of the example was to show the use of automatic verification techniques to check the correctness of a given control logic in order to guarantee safe operation. A faulty logic similar to the one shown in fig. 5.2 was first proposed. Using the model checking techniques and posing the right questions to the model checker, an unsafe cycle was found to be generated by states 4, 5 and 6, causing the accumulation of fuel in the chamber without the presence of flame. Once this error was spotted by the model checker, the logic was modified manually and checked again. No systematic procedure for correcting errors was presented. Here, the ultimate goal is to synthesise a logic controller for the startup, normal operation and shutdown of the burner satisfying the informal specifications introduced in subsection 3.4. These informal static and dynamic specifications were given in tables 3.2 and 3.3 respectively. The avoidance of fuel accumulation is considered as part of the list of specifications. To maintain a pictorially representable model, the on/off button shown in the logic depicted in fig. 5.2 is eliminated.

In the initial state of the system both valves are closed, the igniter is off and the flame has not been detected. The startup is completed when both valves are open, ignition has been achieved by detecting the flame and the igniter is off. The shutdown brings the system to its initial state in preparation for restarting. Emergency operations are covered by other procedures treated in the subsequent example.

5.4.1.1   Step 1.- Process Modelling

The model of the burner was developed in subsection 2.7.1. Elementary models were defined for each one of the four process components: air and fuel valves, igniter and flame detector. Transitions in the air and fuel valves and igniter models are input transitions and therefore controllable, while transitions in the flame detector are system responses and thus uncontrollable. The model is calculated as the *asynchronous product* of the elementary process models. The resultant structure is shown in fig. 2.6. Note that initial/final (*state* 1) and normal operation (*state* 12) *states* are the only marked *states* in the system. The next step in the synthesis procedure is to obtain the set of static and dynamic specification in the required representation. Afterwards, the supervisory structure is constructed. Once in possession of such a structure, a global controller is synthesised.

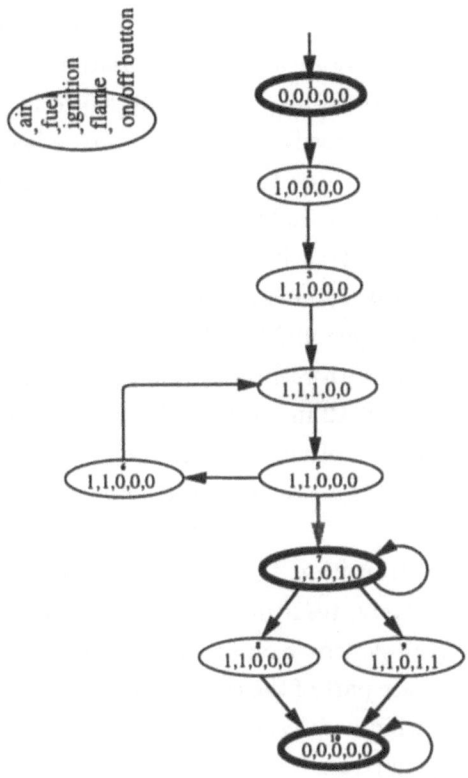

Figure 5.2: Faulty logic proposed for the operation of burner system.

5.4.1.2   Step 2.- Specification Modelling

The translation of specifications in tables 3.2 and 3.3 into PL statements and TL formulas was performed in section 3.4. Forbidden states in table 3.2 are translated into the PL statements shown in table 3.4. Dynamic specifications in table 3.3 are translated into TL formulas 3.12, 3.14 and 3.16 as explained in subsection 3.4.2 and then translated into the *a–machine* domain obtaining the structures shown in figs. 3.2, 3.3 and 3.4.

5.4.1.3   Step 3.- Supervisory Structure Synthesis

The eight forbidden *states* obtained in subsection 3.4 are eliminated from the process model yielding the structure shown in 5.3. A visual check reveals this structure to be uncontrollable because the uncontrollable transition 41 ("flame is detected") can occur in *state* 1 leading to a *state* in which only the flame has been detected (*state* (0, 0, 0, 1) shown in fig. 5.3 in gray line without labelling number) and, while in normal operation (*state* 6), uncontrollable transition

42 can occur unexpectedly, indicating that the flame has been extinguished
leading the system to a *state* in which accumulation of fuel will occur in the
chamber (*state* (1, 1, 0, 0) shown in the same figure in gray line without labelling number). As can be easily seen this structure leads to an empty maximum controllable behaviour due to the existence of uncontrollable transition
41 from *state* 1. Hence, the original Supervisory Control Theory fail to give any
useful results. Considering transitions 41 and 42 as the cause of conditional
controllable behaviour, this same structure can be used as the generator of a
non empty conditionally controllable behaviour. Therefore, this *a−machine* is
conditionally controllable with respect to the process model due to transition
41 in *state* 1 and transition 42 in *state* 6. The supervisory structure is shown
enclosed in the inner square of fig. 5.3 while the *states* causing the system to
be conditionally controllable are shown in the outer square that represents the
process model.

Apart from *state* 1, transition 41 may be executed in *states* 2 or 3, leading
to *states* 5 or 7 respectively, resulting in emergency situations not considered
in the specifications (the system is on fire due to an accident or bad operation)
although *states* 5 and 7 are part of the controllable behaviour on the supervisory
structure. In *state* 6, which represents the normal operation point, the flame
may go off at any moment (transition 42), leading to an emergency situation not
considered either. These emergency situations should be part of the behaviour
covered by the global controller. Alarm devices should be included in the
process to indicate the occurrence of such events. This requires the modification
of the process model and the list of behaviour specifications to consider these
additional cases. This will be shown in the following example.

5.4.1.4   Step 4.- Controller Synthesis

Using the supervisory structure of figure 5.3 as the base model, a local controller is synthesised for each of the three *a−machines* representing the dynamic
specifications D1, D2 and D3 introduced in subsection 3.4.2.

5.4.1.4.1   Specification D1.   Specification D1 (formula 3.12) describes the
startup sequencing of the three controllable process components, air valve, fuel
valve and flame detector. Calculation of the maximum controllable behaviour
with respect to the supervisory structure results again in $K^\uparrow = \emptyset$. Therefore, transitions 41 and 42 are allowed to generate conditionally controllable
behaviour as in the case of the supervisory structure. The result is shown in
fig. 5.4. This *a−machine* is conditionally controllable in *states* 2 and 3 with

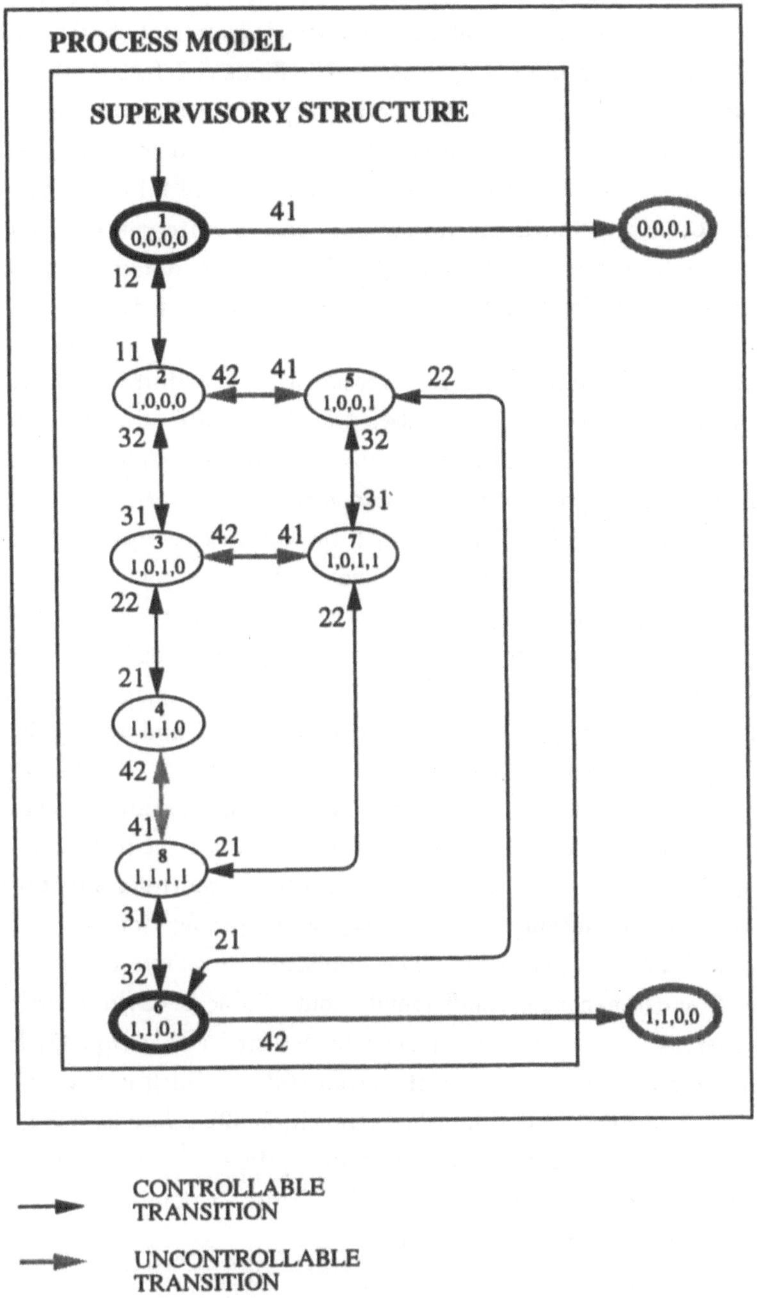

PROCESS MODEL

SUPERVISORY STRUCTURE

Figure 5.3: *a–machine* model after eliminating forbidden states.

respect to the supervisory structure due to the possible occurrence of transition 41. Behaviour after transition 21 reaching *state* 4 is not part of the current specification. Note in the supervisory structure (fig. 5.3) that after the occurrence of transition 21 (fuel valve being opened) from *state* 3, there are only two possible transitions from *state* 4: either the flame is detected (transition 41) or the fuel valve is closed again (transition 22). If the fuel valve is closed, a new path is created to consider this situation and to avoid conflicting with the current specification. This path consists of *states* 5 and 6 in fig. 5.4. From *state* 6 it is possible to reach the initial *state* and execute specification D1 again. Marked *states* correspond to initial/final (*state* 1) and normal operation (*state* 8) points. As can be visually checked this *a–machine* is trim.

5.4.1.4.2   Specification D2.   Specification D2 (formula 3.14) prescribes the exclusive occurrence of only two transitions after the air and fuel valves are open, igniter is on and flame is not yet detected (*state* (1, 1, 1, 0)): either the flame is detected or the fuel valve is closed. The *a–machine* realising the maximum controllable behaviour with respect to the supervisory structure is depicted in fig. 5.5. It is observed that it does not contain any set of *states* in which the specification is totally satisfied. This is because specification D2 forces the system to switch off the igniter once flame has been detected in normal operation (*state* 6 in fig. 5.3). Therefore, the extinguishing of the flame from *state* 8 in fig. 5.3 is an uncontrollable event and *state* 8 is eliminated from the structure to obtain a controllable behaviour. *State* 4 regarding the detection of the flame is also eliminated following the same reasoning. Again, the original Supervisory Control Theory is far too strict to generate useful solutions. If controllability is relaxed to conditional controllability for transitions 41 and 42 (to maintain the consistency with specification D1), the existence of *states* 4 and 8 is permitted. The resultant *a–machine* is shown in fig. 5.6. *States* 5 and 6 generate conditionally controllable behaviour with respect to the supervisory structure. Flame can be detected (transition 41) from *state* 5 leading to an emergency situation. From *state* 6, which is part of the startup sequence, flame can be extinguished (transition 42) before finishing the startup leading to a premature end of operation. Marked *states* correspond to initial/final (*state* 1) and normal operation (*state* 7) points. As in the previous case, the resultant *a–machine* is trim.

5.4.1.4.3   Specification D3.   Specification D3 (formula 3.16) captures the eventual closing of the fuel valve and the flame being extinguished. Calculations for the maximum controllable behaviour of specification D3 with respect to the

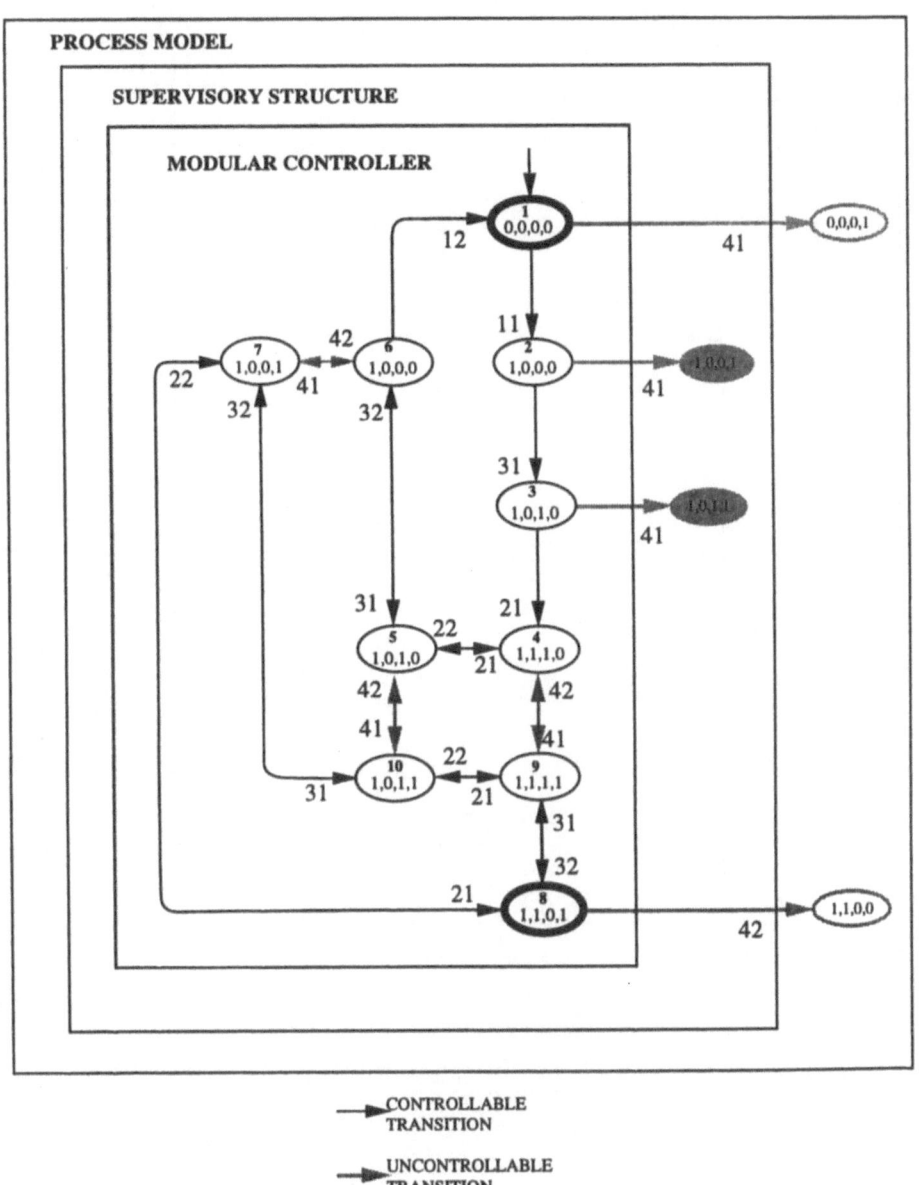

Figure 5.4: Local controller corresponding to specification 1.

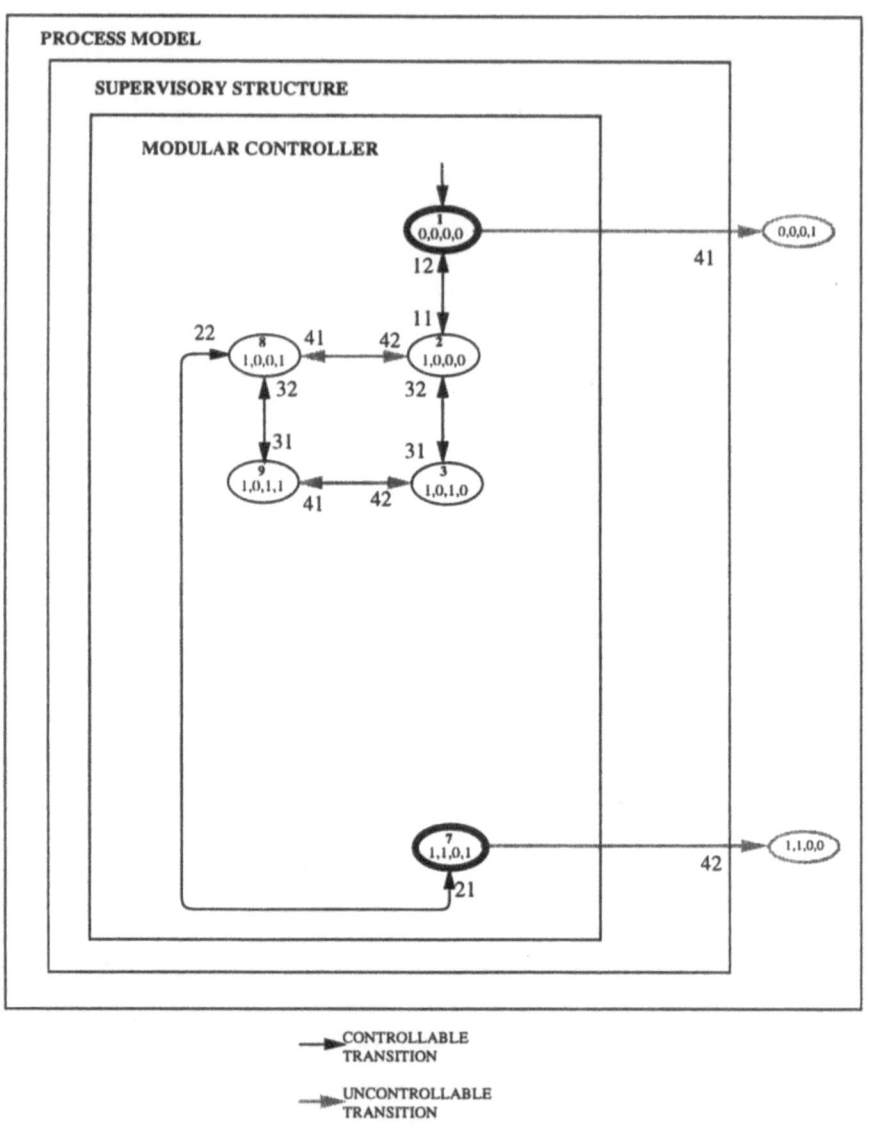

Figure 5.5: Local controller corresponding to specification 2, satisfying the standard controllability definition.

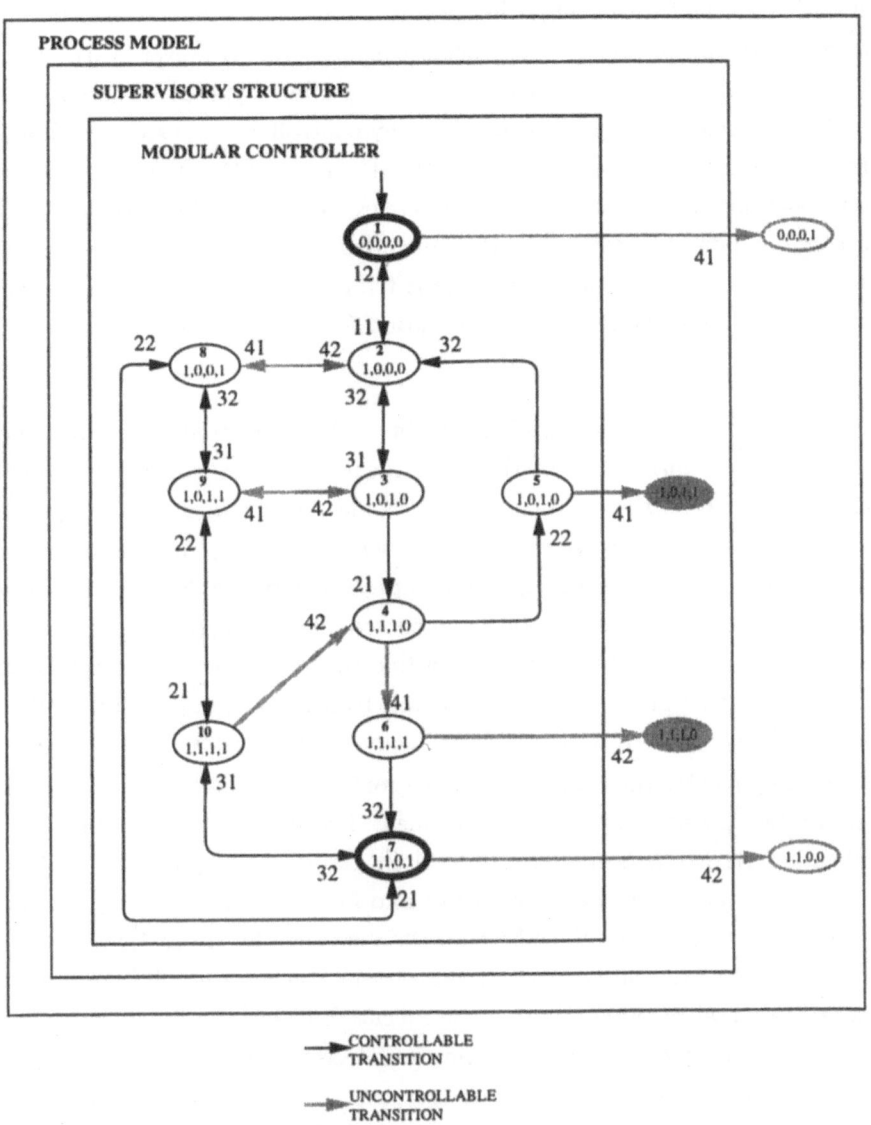

Figure 5.6: Local controller corresponding to specification 2.

supervisory structure result, as in the case of specification D1, in $K^\uparrow = \emptyset$. It can be seen in the supervisory structure of fig. 5.3 that closing the fuel valve (transitions 22) and the eventual flame extinction (transition 42) lead to *state* 2, from where the restart of the process can proceed. However, if the flame is detected again, an emergency situation should be triggered. The only way to avoid this scenario is by not executing any controllable transition ending in *state* 2. This leaves the initial state without controllable transitions and the system with an empty controllable behaviour. Therefore, transitions 41 and 42 are again allowed to cause conditional controllable behaviour. The result is shown in fig. 5.7. Again, the *a–machine* is trim and marked *states* correspond to initial/final (*state* 1) and normal operation (*state* 8) points.

5.4.1.4.4 Global Controller Synthesis.

The global controller is obtained as the *controller conjunction* of the three local controllers previously synthesised. The result is shown in fig. 5.8. Nodes shown in a continuous line represent *states* which the system can reach while in startup, normal operation or shutdown. Nodes shown in a grey line represent undesirable *states* that may be reached after the execution of an uncontrollable transition. Controllable transitions are depicted by a continuous line while uncontrollable transitions are represented by a grey line. Uncontrollable transitions that go to undesirable states are the cause of uncontrollable behaviour in the system.

Inspecting fig. 5.8 the compliance of the controller with all the specifications given in tables 3.2 and 3.3 can be established. The startup sequence defined in specification D1 is represented by *states* 1 (initial state), 2 (air valve open), 3 (air valve open and igniter on) and 4 (air and fuel valves open and igniter on) and associated transitions. Specification D2 is satisfied by *states* 4, 5, 6, 10 and 14 and associated transitions. Once the system reaches *state* 4 from *state* 3, there are only two possible events to be executed: either the flame is detected or the system must restart. If the flame is detected, the system reaches *state* 14 and then the igniter is turned off to reach *state* 10 which is the normal operating point. If the flame is not detected, the fuel valve is closed first, and then the igniter is switched off reaching *state* 6. From this *state*, behaviour not prescribed by any of the specifications can occur. Note that this is originated by the possibility of switching the igniter on or detecting the flame again. In order to eliminate this extra behaviour, a specification must be included to force the system to continue the restart procedure immediately after the igniter has been switched off. Such a specification is described in the next paragraph. Specification D3 describing the shutdown is captured by *states* 10, 11, 12 and 1

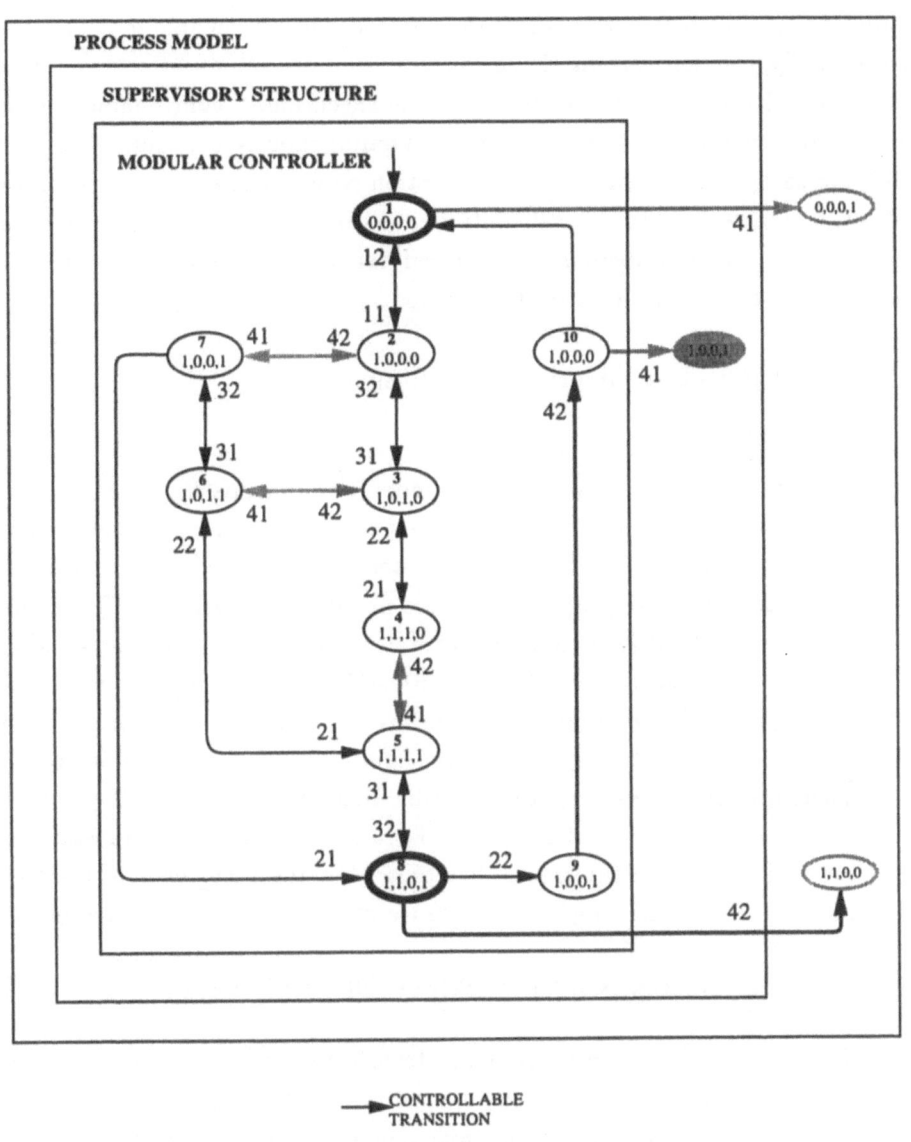

Figure 5.7: Local controller corresponding to specification 3.

and associated transitions. Once the system is in normal operation (*state* 10), only two transitions can occur: either the flame is extinguished (transition 42) leading to an emergency *state* which the controller is not capable to deal with, or eventually, the fuel valve is closed.

Having identified conditions in which the execution of controllable transitions in the closed–loop behaviour is not unambiguously determined, in the following paragraphs the dynamic specification required to eliminate the extra behaviour is given. Then a local controller is calculated for this specification and the *controller conjunction* is performed between this new local controller and the global controller previously obtained.

5.4.1.5   Step 2.- Specification Modelling

- Specification D4.

  The igniter must be switched off under two circumstances, as can be seen in fig. 5.8. The first one is during the startup to cause the system to reach its normal operating point (*state* 10). The second one is when the system is starting up and flame is not detected (*state* 4). Then the system must be driven to its initial state by closing the fuel valve (*state* 5), switching the igniter off (*state* 6) and finally closing the air valve (*state* 1). The generation of extra behaviour commences in *state* 6 from where flame may be detected or the igniter can be switched on again. Therefore, it is necessary to prescribe that after switching the igniter off, the only possible actions allowed are to close either the air valve or the fuel valve. A TL formula that prescribes the occurrence of the two possible transitions after the occurrence of transition 32 is the following.

  $$[(\infty_1, \infty_2, \infty_3, \infty_4) \wedge (\tau = 32)] \rightarrow \bigcirc [(\tau = 12) \, \triangledown \, (\tau = 22)] \qquad (5.1)$$

  The resultant *a–machine* from the translation process is shown in fig. 5.9. Both *states* conforming the *a–machine* are marked because they can potentially represent any of the initial/final or normal operation point of the system. This *a–machine* is trim, as can be visually checked.

5.4.1.6   Step 4.- Controller Synthesis

The local controller for this specification is shown in fig. 5.10. Transition 32 is executed either from *state* 3 or *state* 5. In both cases the only allowed transitions after transition 32 are either transition 12 from *state* 10, as a part of the

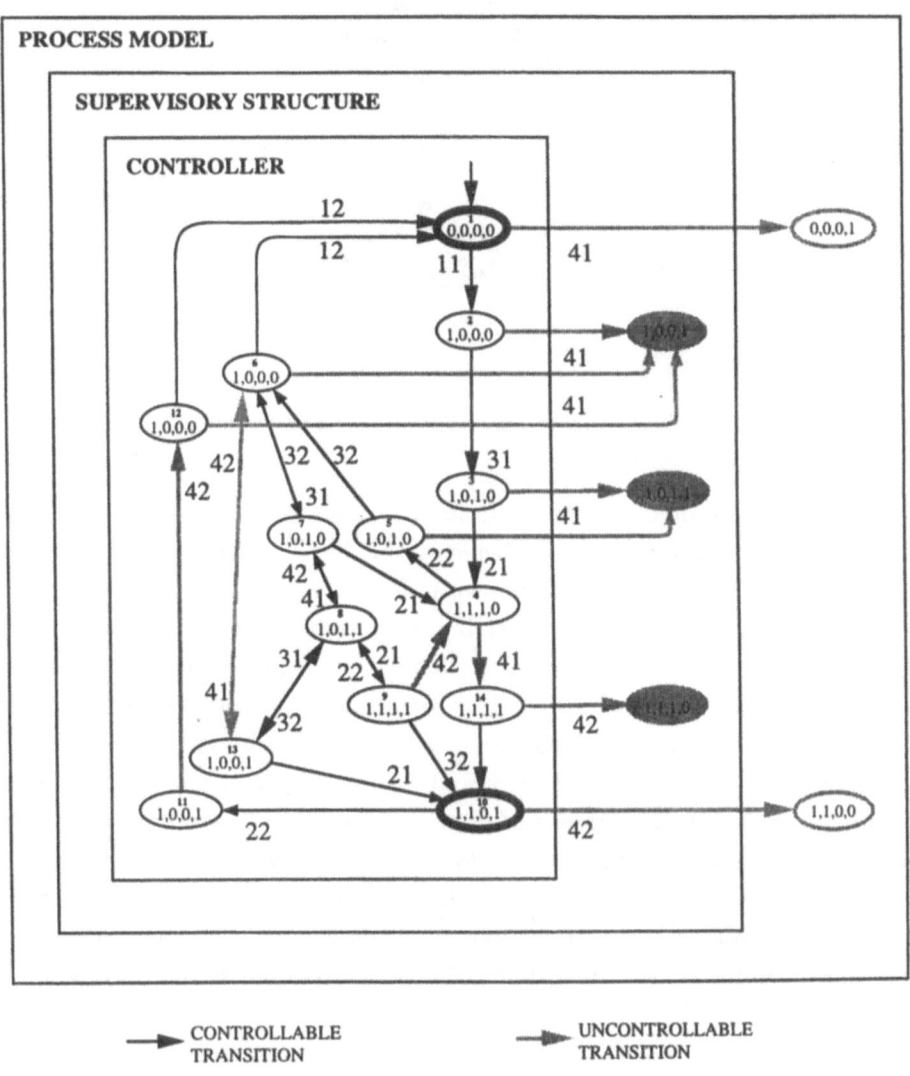

Figure 5.8: *a–machine* corresponding to the global controller for specifications D1, D2 and D3.

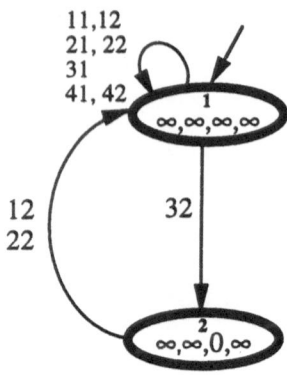

Figure 5.9: *a–machine* corresponding to specification D4.

restart procedure or transition 22 from *state* 7 for the normal shutdown process. The *controller conjunction* between this local controller and the global controller previously obtained is shown in fig. 5.11. All the extra behaviour is eliminated, leaving a structure satisfying the four dynamic specifications. However, the detection or extinction of the flame are uncontrollable events that can occur under many circumstances. This makes the existence of controllable behaviour with respect to the process model conditional upon the no execution of transition 41 from *states* 1, 2, 3 and 5, 6 and 10 and transition 42 from *states* 7 and 8. The occurrence of either transition from the indicated *states* will lead the system to undesirable *states* for which emergency procedures are necessary. A key advantage of the synthesis procedure employed is the systematic identification of *states*/transitions from where emergency operations must be considered. A controller dealing with these situations is the subject of the following section.

5.4.2   Augmented Burner System

The scenarios identified in subsection 5.4.1 with the aid of the notion of conditional controllability in which the system reaches abnormal states, hence requiring emergency intervention, are summarised as follows:

- **Undesirable presence of flame.**

  During startup, there is the possibility of the flame being detected before reaching the state in which flame must be produced (*states* 1, 2 and 3 in fig. 5.11). If the startup process failed and the system is in preparation

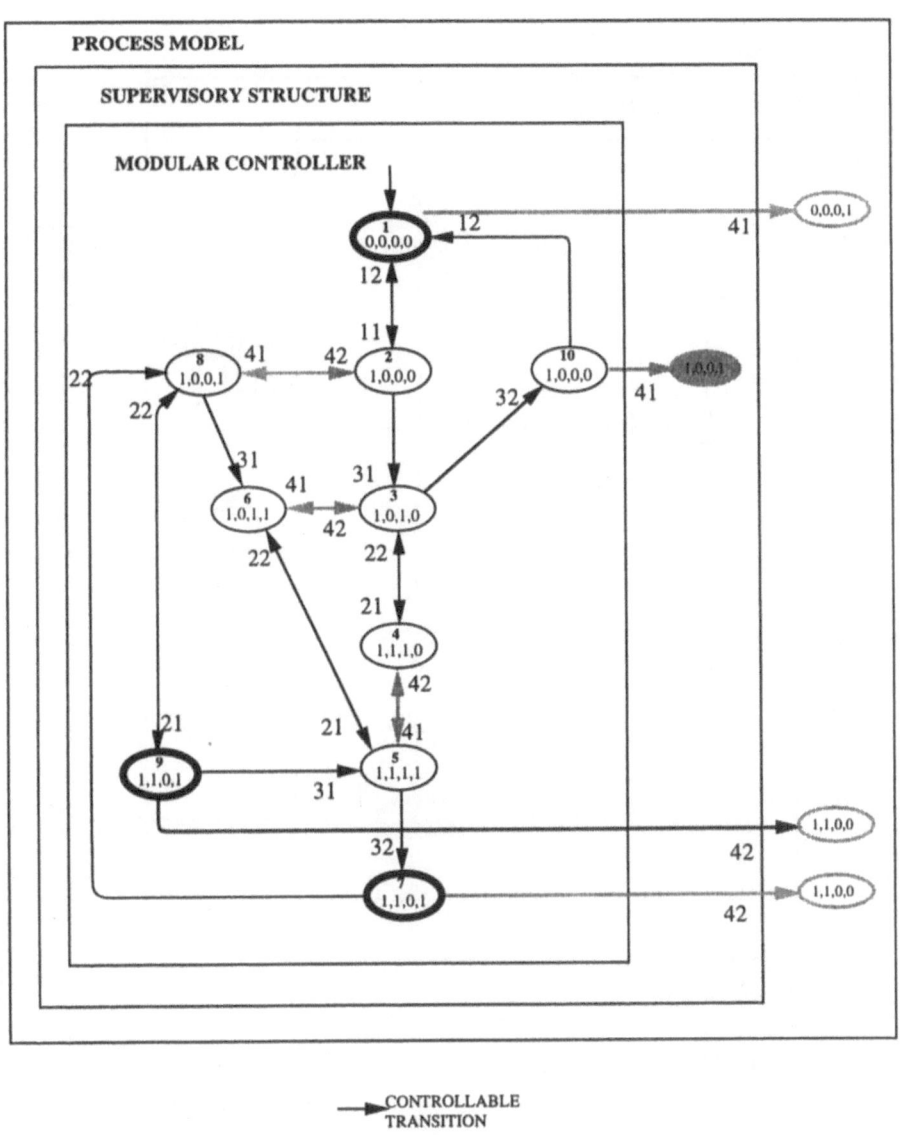

Figure 5.10: *a-machine* corresponding to the local controller for specification D4.

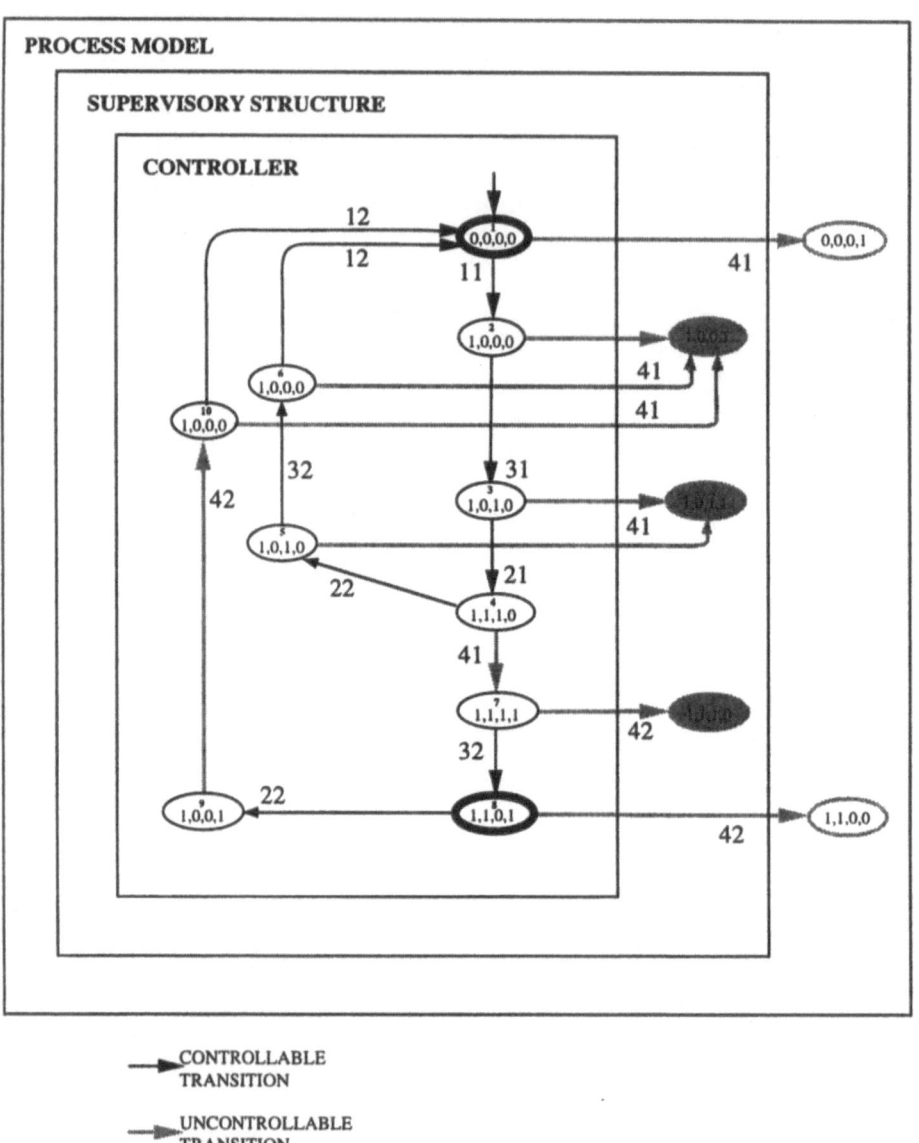

Figure 5.11: *a–machine* corresponding to the global controller for specifications D1, D2, D3 and D4.

for restart, flame may occur from *states* 5 or 6. During the shutdown procedure, once the fuel valve is closed in *state* 9, the next transition that should occur is the flame not being detected. The presence of flame under these circumstances indicates abnormality in the burner operation. Note that the controller of fig. 5.11, does not consider this situation. From the normal operation point (*state* 8), the fuel valve is closed and then the controller assumes that the next transition is the flame extinction. If the flame does not extinguishes, the controller will not take any other action. This is rather a specification problem in which the occurrence of a system response has not been taken into account.

- **Undesirable absence of flame.**

  During the startup process, once the flame has been generated it may be extinguished leading to undesirable situations (*states* 7 and 8 in fig. 5.11) where fuel accumulation will occur.

The task now is to specify and synthesise a controller able to handle these scenarios driving the system to the safest possible states. Considering only the burner system, once an undesirable situation occurs, the safest possible local states are those in which air and fuel valves are closed and the igniter is off. Under any of the circumstances described in the previous paragraph further emergency measures must be taken (i.e. the presence of flame may require an emergency shutdown and to call the fire brigade). Here the system is equipped with alarm devices to be activated in case of emergency and dynamic specifications are included to guarantee that the system reaches the safest local state under emergency conditions. The synthesis procedure goes again through the four proposed steps. The size of the obtained structures precludes their pictorial representation. Thus, all the generated structures, with the exception of the final controller are presented in tabular form in appendix A. In the left hand side of the tables, *states* are listed. *State* 1 is always the initial state of the *a−machine*. For each *state*, columns show if it is marked, its *state−variable* values and what *states* are reached using the transitions listed on the top of each table.

5.4.2.1   Step 1.- Process Modelling

The process model for the extended example is the same as the one used for the burner system but augmented with two alarm devices to indicate the undesirable presence or absence of flame. Each alarm device is modelled as an *a−machine* with two *states*. *State−variable* values and transitions for each alarm

elementary a−machine	state−variables		transitions	
alarm for undesirable flame presence	alarm status	0: off 1: on	51 52	switching on switching off
alarm for undesirable flame absence	alarm status	0: off 1: on	61 62	switching on switching off

Table 5.1: List of states and controllable transitions for the elementary a−machines of the alarm devices.

device are shown in table 5.1. These two new elementary a−machines are added to the previous process model using the asynchronous product, making the process model to grow from 2^4 to 2^6 states. Two new states are added to the list of marked states corresponding to the safe terminal states that the system must reach during emergency operation: when any of the two alarms is active, valves are closed, igniter is off and flame is not detected. The model is presented in tabular format in table A.1. Only four states are marked: the initial state (1), the normal operation point (12) and safe states in which alarm for undesirable flame presence is active (17) and the alarm for undesirable absence of flame is active (33) with the valves closed, the igniter off and no flame detected.

5.4.2.2   Step 2.- Specification Modelling

As mentioned in the previous paragraph, the existence of undesirable process behaviour impose modifications in the static and dynamic specifications used in the previous example. These specifications will be used as the starting point for the determination of those for the augmented system and at the same time they will illustrate alternative ways of constructing specifications. For the handling of abnormal operation and use of alarms, new dynamic specifications are included.

5.4.2.2.1   Static Specifications. The supervisory structure must contain all possible trajectories to be considered by the global controller including emergency operation. This means that the static specifications must describe forbidden states that can be avoided only by the execution of controllable transitions in order to maintain the supervisory structure controllable. To achieve this, the static specifications introduced in table 3.4 and discussed in subsection 3.4.1, are modified, except for specification S1 and statement s_{2a}. Specification S1

Spec S1	No ignition unless necessary	s_1	$(0, \infty_2, 1, \infty_4, \infty_5, \infty_6)$
Spec S2	No fuel unless necessary	s_2	$(0, 1, \infty_3, \infty_4, \infty_5, \infty_6)$

Table 5.2: Forbidden states satisfying the static specifications for the augmented burner system

describes the avoidance of switching the igniter on when the air valve is not open. These circumstances can be avoided by not executing the appropriate controllable transitions. Statement s_{2a} from specification 2 does the same for the fuel valve. It must never be open unless the air valve is open. On the other hand, statement s_{2b} models the *state* in which the air and fuel valves are open without the flame being detected. This situation can occur due to the spontaneous detection of flame (uncontrollable transition 41) and it cannot be avoided by manipulating any controllable transition. This behaviour must be considered within the supervisory structure and therefore, statement s_{2b} must be discarded. On the same grounds, specification S3 is discarded as well. Instead, they must be incorporated into the supervisory structure as a part of handling of the abnormal *states* from where the emergency procedures are initiated. The remaining forbidden states specifications are shown in table 5.2.

5.4.2.2.2 Dynamic Specifications. Dynamic specifications must consider normal and abnormal operation as well as the handling of the alarm devices. The previous example only treated normal operation by using the three dynamic specifications introduced in table 3.3 plus the extra specification discussed in subsection 5.4.1. These specifications are modified and augmented to consider emergency operation during the abnormal presence or absence of flame. The first four specifications presented here, dealing with startup, shutdown and normal operation are based on those used in the previous example. Specifications 5 to 8 prescribe the use of the alarm devices while specifications 9 to 11 deal with the actual emergency operations once either alarm has been activated. All the corresponding *a–machines* of the specifications were checked to be trim. They are shown in tabular form in appendix A.

- **Specification D1. Startup.**

  - Specification D1a. Startup Sequencing.

    Specification D1 in table 3.3 described the normal starting sequence
    of the air and fuel valves and the igniter without considering a spe-
    cific initial state. This forced the sequence to be executed every time
    that the air valve was opened regardless of the status of the rest of
    the components. If the same approach is used in the present case, in
    order to prescribe the same behaviour but considering as well pos-
    sible abnormal conditions, the detection or extinction of the flame
    must be included in every state of the sequence. This is achieved by
    introducing the detection or disappearance of flame (transitions 41
    and 42) as possible transitions to be executed from each state after
    the air valve is open. This behaviour is prescribed for the present
    case in the following TL formula

    $$[(\infty_1, \infty_2, \infty_3, \infty_4, \infty_5, \infty_6) \land (\tau = 11)] \rightarrow$$
    $$\{[\bigcirc(\tau = 41)] \;\nabla\; [\bigcirc(\tau = 42)] \;\nabla\;$$
    $$\{[\bigcirc(\tau = 31)] \rightarrow \bigcirc[(\tau = 41) \;\nabla\; (\tau = 42) \;\nabla\; (\tau = 21)]\}\} \quad (5.2)$$

    The corresponding *a–machine* is shown in tabular form in table A.2.
    The marked *states* correspond to the marked *states* in the process
    model: initial, final and normal operating points as well as safe
    states in case of emergency operation.

  - Specification D1b. Initial System State.

    The use of a completely covered state as a first state in the previous
    specification does not permit to unambiguously define the operating
    starting point of the process, as in the case of the burner example.
    The existence of the alarm devices introduces extra controllable be-
    haviour that is not specified. Using the above specification, alarms
    could be already activated or even the flame may be detected when
    the controller tries to start operation by issuing the control command
    to open the air valve. To avoid this situation, a totally refined state
    is introduced from which the opening of the air valve or detection
    of flame are prescribed as the only possible transitions.

    $$(0, 0, 0, 0, 0, 0) \;\rightarrow\; \bigcirc[(\tau = 41) \;\nabla\; (\tau = 11)] \quad (5.3)$$

    The corresponding *a–machine* is shown in tabular form in table A.4.

- **Specification D2. Reaching Normal Operation Point.**

Specification D2 in table 3.3 and its corresponding TL formula shown in equation 3.14 describe the conditions in which the flame must occur during the startup and what are the actions that must be taken afterwards in order to reach the normal operation point. If both air and fuel valves are open and the igniter is on, then flame may be detected. If not, the controller must initiate the recovery procedure to restart the burner again after failing in detecting the flame. The extinction of the flame must also be considered as a possible event immediately after the flame has been detected during the startup, driving the system to an undesirable *state*. Note that the *state-variable* values would be the same (i.e. $((1, 1, 1, 0, \infty_5, \infty_6))$ for these two different operating conditions. Therefore, because different actions are taken according to which condition occurs, a way of differentiating them must be introduced. This is done by dividing this specification into two parts. In the first part, the reaching of the normal operation point and starting of the recovery procedure are prescribed in the same way as in TL formula 3.14. The second part makes the differentiation between the occurrence of the same *state-variable* values in different *states* by prescribing the behaviour one step back in the operation.

- Specification D2a.

  Again, as in the previous specification, the possible execution of the uncontrollable transitions after the detection of the flame must be incorporated in the TL formula. Using TL formula 3.14 as a basis, the flame extinction (transition 42) is included as follows

$$(1, 1, 1, 0, \infty_5, \infty_6) \rightarrow$$
$$[\{[\bigcirc(\tau = 41)] \rightarrow \bigcirc[(\tau = 32) \, \triangledown \, (\tau = 42)]\} \, \triangledown$$
$$\{[\bigcirc(\tau = 22)] \rightarrow \bigcirc[(\tau = 32) \, \triangledown \, (\tau = 42)]\}] \qquad (5.4)$$

  The corresponding *a-machine* is shown in tabular form in table A.3.

- Specification D2b.

  As mentioned previously, *state* $(1, 1, 1, 0, \infty_5, \infty_6)$ can be part of the normal startup sequence (after switching the igniter on and opening the fuel valve) or it can be an undesirable state (after the unexpected extinction of the flame during startup) from which an emergency procedure must commence. In the latter, the alarm for the undesirable absence of flame will be activated. Specification D2a does not distinguish between these two situations. A TL formula is added to make this distinction by prescribing the behaviour of the startup one step before the occurrence of *state* $(1, 1, 1, 0, \infty_5, \infty_6)$, that is, before opening the fuel valve:

$$[(1, 0, 1, 0, 0, 0) \wedge (\tau = 21)] \rightarrow \bigcirc[(\tau = 22) \triangledown (\tau = 41)] \qquad (5.5)$$

TL formula 5.5 guarantees that after opening fuel valve during normal operation, the only executable actions are the detection of the flame or the closing of the fuel valve and therefore none of the alarms will be activated. The corresponding *a–machine* is shown in tabular form in table A.5.

- **Specification D3. Shutdown.**

Specification D3 in table 3.3 describing the shutdown must be modified as well. Instead of prescribing only the shutdown from the normal operation point, the unwanted presence of flame must also be included. Thus, from the normal operation point, two situations are considered:

  – The flame is extinguished causing an undesirable situation.

  – The shutdown starts by closing the fuel valve.

If the shutdown is initiated by closing the fuel valve and flame does not disappear, then the alarm for undesired presence of flame must be activated. On the other hand, if the flame is extinguished the shutdown can close the air valve but always considering the possibility of the flame being detected again. This is reflected in the following formula:

$$(1, 1, 0, 1, 0, 0) \rightarrow$$
$$\{[\bigcirc(\tau = 42)] \triangledown \{[\bigcirc(\tau = 22)] \rightarrow \{[\bigcirc(\tau = 51)] \triangledown$$
$$\{[\bigcirc(\tau = 42)] \rightarrow \bigcirc[(\tau = 12) \triangledown \tau = 41)]\}\}\} \qquad (5.6)$$

The corresponding *a–machine* is shown in tabular form in table A.6.

- **Specification D4. Avoiding Undesirable Restart of Igniter.**

Specification D4 introduced in subsection 5.4.1 to avoid the undesirable operation of the igniter must be modified as well. It can be seen in fig. 5.11 that after the occurrence of transition 32 from *state* 7, transitions 22 or 42 may occur. In the case of transition 32 from *state* 5 reaching *state*

6, not only transition 12 but transition 41 may also occur. This results in the following formula:

$$[(\infty_1, \infty_2, \infty_3, \infty_4, \infty_5, \infty_6) \wedge (\tau = 32)] \rightarrow$$
$$\bigcirc[(\tau = 12) \nabla (\tau = 22) \nabla (\tau = 41) \nabla (\tau = 42)] \qquad (5.7)$$

The corresponding *a–machine* is shown in tabular form in table A.7.

The handling of undesirable operation states imposes specifications for emergency procedures and the use of the alarm devices. As mentioned in subsection 5.4.2.1, the controller uses alarms to indicate the undesirable presence or absence of flame in the chamber. Dynamic specifications 5 to 8 prescribe the operation of these alarm devices.

- **Specification D5. Activation of Alarm for Undesirable Flame Presence.**

  Flame detection can be undesirable during either startup or shutdown. The circumstances in which this occurs are different as explained in the following description and must be prescribed by different TL formulas.

  - Specification D5a. Undesirable presence of flame during startup.

    The flame is undesirable when it occurs before the system reaches the proper conditions for flame generation. The opening of the fuel valve is the last action before the occurrence of flame during the startup. If the flame is detected before this event, the alarm for undesirable flame must be triggered immediately.

    $$[(\infty_1, 0, \infty_3, 0, 0, 0) \wedge (\tau = 41)] \rightarrow \bigcirc(\tau = 51) \qquad (5.8)$$

    The corresponding *a–machine* is shown in tabular form in table A.8.

  - Specification D5b. Undesirable presence of flame during shutdown.

    When the fuel valve closes during the shutdown process, flame must be extinguished afterwards. If this does not occur, the system enters abnormal operation thereby triggering the alarm for undesirable presence of flame.

    $$[(1, 1, 0, 1, 0, 0) \wedge (\tau = 22)] \rightarrow \bigcirc[(\tau = 51) \nabla (\tau = 42)] \qquad (5.9)$$

    The corresponding *a–machine* is shown in tabular form in table A.9. Note that the simpler TL formula $(1, 0, 0, 1, 0, 0) \rightarrow \bigcirc[(\tau = 51) \nabla (\tau = 42)]$ can not be used because state $(1, 0, 0, 1, 0, 0)$ may also occur as an abnormal state during startup after the air valve is opened, as can be observed in fig 5.8.

- **Specification D6. Activation of Alarm for Undesirable Flame Absence.**

  Once the flame has been detected during the startup, it should not be extinguished until the fuel valve closes during the shutdown. If the flame disappears under any other circumstance, the alarm for undesirable absence of flame must be activated:

  $$[(1, 1, \infty_3, 1, 0, 0) \wedge (\tau = 42)] \rightarrow \bigcirc(\tau = 61) \tag{5.10}$$

  The *state-variable* corresponding to the igniter status is covered because the flame can be extinguished even when it has just been detected during the startup and the igniter has not been switched off yet. The corresponding *a-machine* is shown in tabular form in table A.10.

- **Specification D7. Use of Alarm for Undesirable Presence of Flame.**

  The use of the alarms must be restricted only to emergency or abnormal conditions. Therefore if flame is not present, the alarm for undesirable presence of flame must not be activated under any circumstance:

  $$(\infty_1, \infty_2, \infty_3, 0, \infty_5, \infty_6) \rightarrow \bigcirc(\tau \neq 51) \tag{5.11}$$

  The corresponding *a-machine* is shown in tabular form in tables A.11.

- **Specification D8. Use of Alarm for Undesirable Absence of Flame.**

  - Specification 8a.

    If flame is present then the alarm for undesirable absence of flame must never be activated.

    $$(\infty_1, \infty_2, \infty_3, 1, \infty_5, \infty_6) \rightarrow \bigcirc(\tau \neq 61) \tag{5.12}$$

  - Specification 8b.

    On the other hand, if the flame is not present and the fuel valve is closed, then the absence of flame is part of the normal behaviour and therefore the alarm must no be activated.

    $$(\infty_1, 0, \infty_3, 0, \infty_5, \infty_6) \rightarrow \bigcirc(\tau \neq 61) \tag{5.13}$$

The corresponding *a–machines* are shown in tabular form in tables A.12 and A.13.

Dynamic specifications D9, D10 and D11 deal with the actual emergency procedures to be executed once an abnormal condition occurs and either alarm has been triggered. Two types of specifications are presented: those to prescribe what must not be done during emergency operation (i.e. to avoid worsening the emergency condition) and those to prescribe what to do in order to drive the system to a safe state.

- **Specification D9. Prohibited actions.**

  Once either alarm is activated, normal operation is abandoned and the system must be driven to a safe state. The first step is to prohibit any action that can worsen the situation. That is, opening any of the closed valves, switching the igniter on, disabling the activated alarm or activating the wrong alarm. Formulas for each of the alarms are presented separately.

  - Specification D9a.

    If the alarm for undesirable flame presence is activated, TL formula 5.14 captures the required behaviour.

$$(\infty_1, \infty_2, \infty_3, \infty_4, 1, \infty_6) \rightarrow$$
$$\Box[\ (\tau \neq 11) \land (\tau \neq 21) \land (\tau \neq 31) \land (\tau \neq 52) \land$$
$$(\tau \neq 61) \land (\tau \neq 62)] \tag{5.14}$$

  The corresponding *a–machine* is shown in tabular form in table A.14.

  - Specification D9b. The corresponding behaviour when the alarm for the undesirable absence of flame is activated is prescribed by the following TL formula:

$$(\infty_1, \infty_2, \infty_3, \infty_4, \infty_5, 1) \rightarrow$$
$$\Box[\ (\tau \neq 11) \land (\tau \neq 21) \land (\tau \neq 31) \land (\tau \neq 51) \land$$
$$(\tau \neq 52) \land (\tau \neq 62)] \tag{5.15}$$

  The corresponding *a–machine* is shown in tabular form in table A.15.

- **Specification D10. Driving the system to a safe state when alarm for undesirable flame presence has been activated.**

As can be observed in fig. 5.11 the air valve may be open or the igniter may be on or both occur simultaneously when the alarm for undesirable presence of flame has been triggered. The appropriate emergency procedure must drive the system to a safe state executing actions in a sequential fashion.

- Specification D10a.

  In order to minimise the wear of the igniter, if it is on, it must always be switched off at the first instance before closing the air valve:

$$(\infty_1, 0, 1, 1, 1, 0) \rightarrow \bigcirc(\tau = 32) \tag{5.16}$$

- Specification 10b.

  Once it is guaranteed that the fuel valve is closed and the igniter is off, if the air valve is still open, it must be closed:

$$(1, 0, 0, 1, 1, 0) \rightarrow \bigcirc(\tau = 12) \tag{5.17}$$

  The corresponding *a-machines* are shown in tabular form in tables A.16 and A.17.

- **Specification D11. Emergency procedure to drive the system to a safe state when the alarm for undesirable flame absence has been activated.**

  If the alarm for undesirable absence of flame is activated, a similar situation as described in specification D10 exists, but in this case the fuel valve may be open. Therefore the sequence that drives the system to a safe state must impose to close the fuel valve first and then continue with with a procedure equal as the one proposed in specification D10.

  - Specification D11a.

    If the alarm for undesirable flame absence has been activated and the fuel valve is open, it must be closed immediately.

$$(\infty_1, 1, \infty_3, \infty_4, 0, 1) \rightarrow \bigcirc(\tau = 22) \tag{5.18}$$

  - Specification D11b.

    Afterwards, if the igniter is on, it must be switched off.

$$(\infty_1, 0, 1, \infty_4, 0, 1) \rightarrow \bigcirc(\tau = 32) \tag{5.19}$$

– Specification D11c.

Finally, if the air valve is on, it must be closed.

$$(1, 0, 0, \infty_4, 0, 1) \rightarrow \bigcirc(\tau = 12) \tag{5.20}$$

The corresponding *a–machines* are shown in tabular form in tables A.18, A.19 and A.20.

5.4.2.3   Step 3.- Supervisory Structure Synthesis

The PL statements in table 5.2 generate 24 forbidden *states*. They are eliminated from the *a–machine* modelling the process (table A.1). The resultant structure is tested positively for controllability. It is shown in table A.21.

5.4.2.4   Step 4.- Controller Synthesis

Local controllers are synthesised for each of the 19 dynamic specifications presented. Dynamic specifications 1a, 1b, 2a, 2b and 3 prescribing normal startup and shutdown procedures generate local controllers yielding controllable closed–loop behaviour. It means that these dynamic specifications consider all the possible events occurring in the system. The same occurred for dynamic specifications 7, 8a, 8b, 9a and 9b prescribing the conditions for triggering the alarms and the restrictions in their use. On the other hand, dynamic specifications 4 and 5 (to activate the alarms), specifications 10a and 10b (emergency procedures for undesirable flame presence) and specifications 11a, 11b and 11c (emergency procedures for undesirable absence of flame), give rise to local controllers whose behaviour is conditionally controllable with respect to transitions 41 (flame detection) and 42 (flame extinction). This is because if an abnormal situation occurs and the corresponding alarm is activated, the controller must drive the system to a safe state, regardless of the existence of other uncontrollable transitions along the path or even if the cause of the emergency disappears (i.e. the detection of flame presence or absence). Tables A.22, A.23, A.24, A.25, A.26, A.27 and A.28 show the local controllers corresponding to the specifications listed above whose closed–loop behaviour is conditionally controllable. The *states* originating conditional controllability are indicated together with the uncontrollable transitions causing it.

The global controller is given by the *controller conjunction* of the 19 local controllers and is shown in fig. 5.12. Three regions are identified. Region A consisting of 10 *states* covers startup (*states* 1, 2, 3, 4, 5, 6, 14 and 15), shutdown (*states* 15, 16, 17 and 1) and normal operation (*state* 15). This corresponds to the controller synthesised in the previous example and shown in fig. 5.11. As expected, all the possible occurrence of uncontrollable transitions driving the system to abnormal conditions, which were identified in the previous example, are included in regions B and C. Region B considers emergency operation when undesirable flame is detected in the system. *State* 10 is the final safe state where alarm for undesirable flame is on, the valves are closed and the igniter is off. Note that there exists a transition activating the alarm from *state* 16, which is part of the shutdown procedure, as stated in dynamic specification D4a. Region C corresponds to the emergency procedures when there is an undesirable absence of flame. In this case, the system can reach *states* 22 or 28, which are safe.

5.5   Summary

Supervisory Control Theory only provides control mechanisms which maintain the closed–loop process behaviour within a finite set of controllable trajectories by disallowing controllable transitions. This control mechanism is suitable for the compliance of specifications that require the avoidance of behaviour such as forbidden states, but is not able to enforce the execution of controllable transitions. This requires a mechanism capable of determining the necessary control actions to be executed in the controlled process. To achieve this, two new definitions were introduced:

- *Local controller*, a structure describing the enforcement of behaviour pre- scribed by a dynamic specification.

- *Global controller* as the *controller conjunction* of all the *local controllers*, mapping each system state into at most one controllable transition.

The synthesis of either a local controller or a supervisor is equivalent to the synthesis of a controllable or conditionally controllable language as defined in chapter 4. Therefore, in terms of synthesis of languages, there is no distinction between "supervisor" or "controller". However a "controller" is interpreted as realising a language defined by dynamic specifications, while a "supervisor" does the same for forbidden states.

A local controller language is defined as the intersection of the languages realised by the local controller C and the supervisory structure S in which only strings from the supervisory structure that are specified by the controller survive. In this way it is guaranteed that the local controller will generate a closed–loop language that not only satisfies the given dynamic specification but also avoids all the forbidden states.

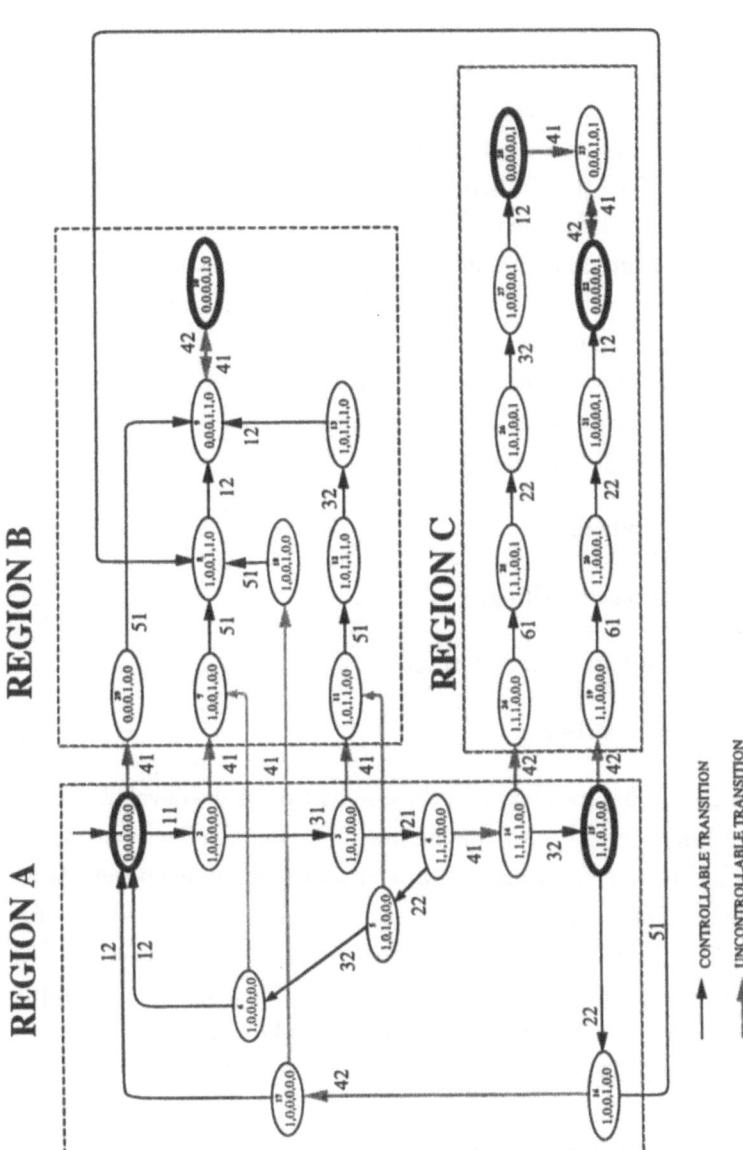

Figure 5.12: α-*machine* corresponding to the final logic/sequential controller for the augmented burner system.

A local controller is guaranteed to be nonblocking, and therefore *proper* in the sense of proposition 4.1, if $\overline{L}_m(C)$ and $\overline{L}_m(S)$ are nonconflicting. The supervisory structure was defined as trim $(L(S) = \overline{L}_m(S))$ and thus, prefix-closed $(L(S) = \overline{L}(S))$. Thus, if the controller is trim and the marked states in both *a–machines* match, the existence of a *proper* controller is guaranteed (see theorem 4.1).

Again, according to proposition 4.2, in order to guarantee the existence of a *global controller*, the marked languages of all *local controllers* in the conjunction must be nonconflicting. If a *global controller* contains *states* in which more than one controllable transition exist, extra dynamic specifications must be included in order to refine the desired behaviour and eliminate all but one controllable transitions. This is achieved by generating *local controllers* and doing the *controller conjunction* with the existing global controller. At present no method is proposed to carry out this specification refinement process, leaving it entirely to the user's judgement.

Using the tools developed in chapter 2, 3 and 4 a controller synthesis procedure has been implemented based on an input–output interpretation of the process. The procedure is divided into four steps:

- Step 1.- Process modelling.

- Step 2.- Specification modelling.

- Step 3.- Supervisory structure synthesis.

- Step 4.- Controller synthesis.

The use of the synthesis procedure and the application of the developed tools were demonstrated in two examples. In the first one, a controller was synthesised for the burner system for which its model and behaviour specifications were developed in section 2.7.1. Given a set of forbidden states, an *a–machine* representing the supervisory structure was synthesised. It was shown that the use of controllability as defined in Supervisory Control Theory yields an empty language. Conditional controllability is then used to allow in the supervisory structure the existence of *states* that must be visited during normal operation but are potential originators of abnormal behaviour. Based on this structure, local controllers are synthesised for each one of the dynamic specifications. The execution of the *controller conjunction* gives rise to a structure that does not comply with the definition of the global controller. Although it satisfies all the given dynamic specification, the structure gives rise to extra behaviour whereby under certain circumstances, more than one controllable transition can be executed. A dynamic specification is introduced to eliminate the undesirable extra behaviour.

The synthesis procedure utilised is capable of systematically highlighting all the sources of uncontrollable behaviour in the process model. This avoids the risk of overlooking necessary specifications for the controller synthesis. How to deal with each source of uncontrollable behaviour is still an engineering task not considered in this work.

Having identified in the first example the possible sources of abnormal behaviour by using conditional controllability, the second example deals with the synthesis of a controller able to handle abnormal operation. The model was modified to handle abnormal behaviour and extra specifications were included. The obtained controller was shown pictorially. Its implementation as a controller requires time–related information that the a–$machine$ structure does not posses. For instance, it has been assumed that if a controllable transition must be executed from a given $state$, this must be prescribed by a specification statement. This will be reflected in the procedural controller as a $state$ from which only the desired controllable transition is executed. If the same controllable transition appears in other $states$ together with uncontrollable transitions, it means that such a behaviour was not relevant to none of the specifications and thus any transition can occur. Supervisory Control Theory does not give any criteria to decide whether the controller must execute the controllable transition first or must wait for a response from the plant. This is an important limitation of the FSM modelling framework and the Supervisory Control Theory which will be explored in the following chapter.

Chapter 6

Implementation Issues

6.1 Introduction

In the previous chapter, an input–output control model was proposed to differentiate between the roles of process and procedural controller as well as between transitions that are process responses (uncontrollable transitions) issued by the process and accepted by the controller and control commands (controllable transitions) that must be issued by the controller. The notion of a *global controller* was also introduced as an *a–machine* in which at most one controllable transition is associated with each *state* together with a finite number of uncontrollable transitions. The tools developed in chapters 2, 3 and 4 were incorporated in a synthesis method and its use was demonstrated in two simple examples. The conditional controllability concept proved to be very useful in the construction of the procedural controller by identifying uncontrollable transitions from those conditions in which the prescription of a controllable transition is required to fulfill behaviour specifications (i.e. for those *states* in which a control action must be forced to occur before anything else happens). However, if more than one transition is associated with a *state*, the controller does not have information to decide whether it must wait for a process response to occur or to execute a control command. This is an inherent limitation originated in the *FSM* framework used to model the controller. In this chapter, this limitation is overcome by assuming that when having more than one transition from a *state* in a procedural controller structure, process responses (uncontrollable transitions) have execution priority over control commands (controllable transitions). However, if only one control command exists in a controller state, is because its execution was prescribed by a dynamic specification and it must be issued to the controlled process at once. Global controllers are synthesised for two examples with "continuous–time" dynamics modelled by differential equations. When there exists several uncontrollable transitions in a given controller *state* with one or less controllable transition, the controller must await a fixed period of time, called *sampling time*, for a uncontrollable transition to occur. If after this period of time no uncontrollable transitions has occurred, the controllable transition, if present, is executed. The process models, together with the corresponding synthesised controllers are implemented in a suitable general purpose dynamic simulator to demonstrate the controllers performance during process operation. *gPROMS* simulator (Barton and Pantelides, 1994) was chosen for this task because it considers the execution of discrete actions in an explicit fashion making the appropriate structural changes in the mathematical model. The simulation is driven by a *FSM*–like command structure according to the occurrence of time or state events. The first example is the buffer system for which a discrete–event model was developed in subsection 2.7.2. A controller is synthesised to fill and drain the buffer tank cyclically. No operation other than the normal mode is considered. As in the burner example, the synthesis procedure is followed graphically step by step. In this example, as in the examples of chapter 5, uncontrollable transitions are generated only by one elementary process component. Therefore, there exists at most one uncontrollable transition from each controller *state*. A typical simulation shows how the controller drives cyclically the tank level from empty to full. The second example consists of a metering system formed by a tank, a level indicator and a set of three valves to fill and drain the tank. The possibility of failure in two of the valves is modelled. As opposed to the other three examples, there exists more than one elementary component capable of producing responses (uncontrollable transitions) from each system *state*. Thus, it is possible to have several uncontrollable transitions from the same controller *state*. The synthesised controller covers in detail the whole operation from startup, normal operation, shutdown and abnormal conditions. As in the case of the surge tank, models of the metering system and the resultant controller are implemented in the dynamic simulator. The results of a typical simulation are shown to demonstrate the controller performance in driving the operation.

6.2   Buffer System

The first example of this chapter is the buffer system introduced in subsection 2.7.2 and depicted there in fig. 2.7. It consists of a tank and a level measuring device with three switches to indicate when the liquid level is minimum, normal level or maximum. Liquid is fed at constant rate into the tank using on/off valve FV and is discharged through on/off valve DV at the bottom of the tank. It is assumed that both valves operate perfectly, i.e. they open or close without failing. A controller is required to fill the tank up to the normal level and empty it cyclically. The initial state of the system is defined with both valves closed and the level measurement indicating minimum. Operation starts by opening feed valve FV which is maintained in this position until the level measurement indicates normal. Then, FV is closed and DV is opened. DV remains open until the level measurement indicates minimum. At this stage DV is closed and FV is opened again to repeat the cycle. The operation is executed indefinitely. Level maximum indicates that the tank is at its maximum capacity and a spill will occur unless corrective actions are taken. No forbidden specifications are given in advance. Therefore, the controller is synthesised following the procedure proposed in chapter 5 but skipping step 3 of the procedure, thus using the process model as the supervisory structure.

6.2.1   Step 1.- Process Modelling

The process model was constructed in subsection 2.7.2 using both valves and the level measurement device as elementary components. First, models for each elementary component were developed. Each of the on/off valves was modelled as a two–*states a–machine* while the level measurement device was modelled as an *a–machine* with three *states*, each representing the level measurement value currently active. The elementary models were presented in table 2.2. The set of transitions of each elementary components are disjoint. Therefore, a model of the process was obtained by the *asynchronous product* of the three elementary models following the same procedure as in the process model construction of the burner example in subsection 2.7.1. The obtained structure formed by 12 *states* was shown in fig 2.9. The order of the elementary *state–variables* in each *state* is as follows:

(level measurement status, fill valve position, discharge valve position)

Transitions 11, 12, 13 and 14 corresponding to the change of level measurement are uncontrollable. It is assumed that the only relevant *state* to be marked is the *state* in which the level measurement indicates zero and both valves are closed (i.e. the initial/final operation point). Physically possible changes in the level measurement were prescribed using *a–machines*. These *a–machines* were then used to obtain a simplified model, reducing the number of transitions per-

mitted to occur from each *state*. The construction of the *a—machines* was based
on the intuitive understanding of the process. Here, the same *a—machines* are
systematically constructed by first writing the desired behaviour as TL formulas and then translating them into *a—machines* using the tools developed in
chapter 3. In the following paragraphs the discussion presented in subsection
2.7.2 is revisited with the results presented as TL formulas.

If both valves are closed, the level measurement will keep the same value
until either of the valves is open. Hence, transitions regarding the change
of level measurement cannot physically occur from this state unless there is
a failure in the level measuring device, a circumstance that is not presently
considered. The only transitions that can be executed from this state are those
which open either valve. This is reflected in the following TL formula

$$(\infty_1, 0, 0) \rightarrow \bigcirc [(\tau = 21) \nabla (\tau = 31)] \tag{6.1}$$

If FV is open and DV is closed, the level will increase until reaching the
maximum possible level where the liquid will spill out from the tank and the
level measurement indicates maximum. Also, either FV can be closed or DV
can be opened. The following TL formula represents such behaviour

$$(\infty_1, 1, 0) \rightarrow \bigcirc [(\tau = 11) \nabla (\tau = 12) \nabla (\tau = 22) \nabla (\tau = 31)] \tag{6.2}$$

On the other hand, if DV is open and FV is closed the level measurement
will decrease until reaching the minimum. Also, either FV is allowed to open
or DV to close. Such behaviour is captured in the following formula

$$(\infty_1, 0, 1) \rightarrow \bigcirc [(\tau = 13) \nabla (\tau = 14) \nabla (\tau = 21) \nabla (\tau = 32)] \tag{6.3}$$

If both valves are open, the level may go up or down depending on its current
height and flowrates. No TL formula describes such a detailed behaviour.

These TL formulas, which model the physically possible behaviour, are
translated into their respective *a—machines* which are shown in figures 2.10, 2.11
and 2.12. Then, the *synchronous product* of these *a—machines* and the structure
representing the *asynchronous product* of the elementary process components
(fig. 2.9) is performed. In the resultant *a—machine* only the behaviour common
to all the considered *a—machine* survives. The final structure is shown in figure
2.13.

6.2.2   Step 2.- Specification Modelling

The objective of the controller is to maintain the cyclic operation of the buffer
tank. In this example, no forbidden states are specified in order to show how

dynamic specifications can discriminate among *states* when performing synchronous products.

6.2.2.1   Dynamic Specifications

The dynamic specifications are extracted from the description of the operation as given in the introduction of section 6.2. The operation is divided into four phases:

- **D1.- Start operation.**

- **D2.- Change from filling to emptying.**

- **D3.- Change from emptying to filling.**

- **D4.- Loop operation of valves FV and DV.**

Each one of these phases is captured by one or several TL formulas which are then translated into *a–machines* with the appropriate marked *states*, using the tools developed in chapter 3. A detailed description of each phase follows.

- **D1.- Start operation.**

  The initial state of the system was defined as level measurement in minimum and both valves closed. To start the filling, valve FV is opened. The above is expressed in the following formula

$$(0,0,0) \rightarrow \bigcirc(\tau = 21) \tag{6.4}$$

  The corresponding *a–machine* is shown in fig. 6.1. Note that the specification is satisfied within *states* 1 and 2. From *state* 1 the only output transition is the opening of FV (transition 21).

- **D2.- Changing from filling to emptying.**

  The level will increase to overfull if the controller does not take any action once normal level has been detected. To avoid this situation, FV is closed immediately after normal level is detected. Afterwards, DV is open to start emptying the tank.

$$(1,1,\infty_3) \rightarrow \bigcirc(\tau = 22) \rightarrow \bigcirc(\tau = 31) \tag{6.5}$$

  Note that *state–variable* 3 is covered in order to minimise the size of the resulting structure. Figure 6.2 represents the corresponding *a–machine*

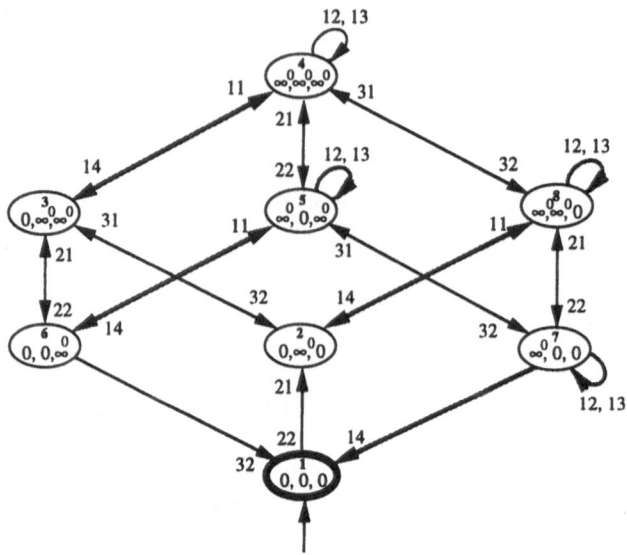

Figure 6.1: *a–machine* corresponding to dynamic specification D1.

for this TL formula. Note that from *state* 3, which is the first state in TL formula 6.5, only the closing of FV is considered, followed by the execution of transition 31.

- **D3.- Changing from emptying to filling.**

  Eventually the tank will be empty and DV must be closed. To start the operation cycle again valve FV is opened. This sequence is covered by the following TL formula

  $$(0, \infty_2, 1) \rightarrow \bigcirc(\tau = 32) \rightarrow \bigcirc(\tau = 21) \tag{6.6}$$

  The corresponding *a–machine* is shown in fig 6.3. As in the previous case, only the relevant *state–variables* are given refined values. The specification is satisfied within *states* 1, 3 and 4 with the sequencing of transitions 32 and 21.

- **D4.-Loop operation of valves FV and DV.**

  None of the TL formulas presented above guarantee that the next transition occurring will be the change of level once one of the valves is open and the other is closed. Thus, a situation may exist in which the controller forces the valves to close and open continuously in an infinite loop. To avoid this situation, TL formulas are introduced to disallow the immediate closing or opening of valves when they have been just opened or

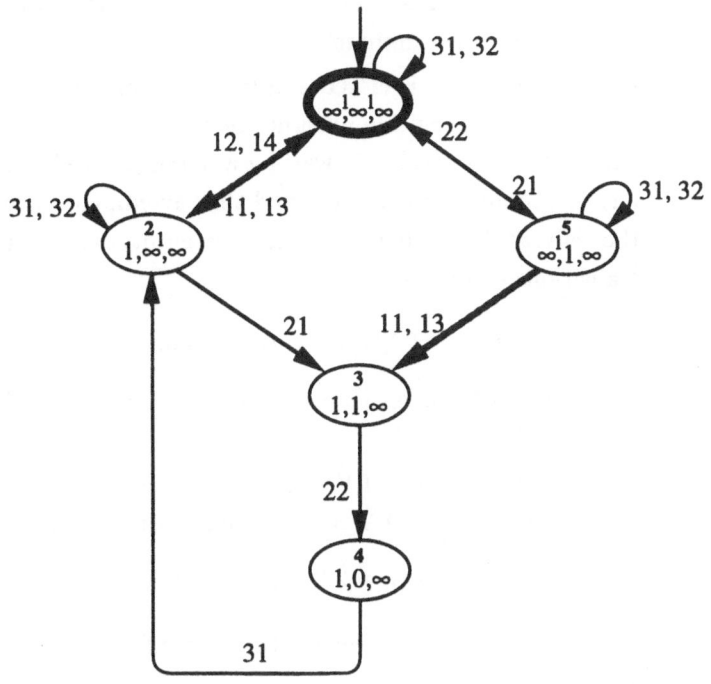

Figure 6.2: *a−machine* corresponding to dynamic specification D2.

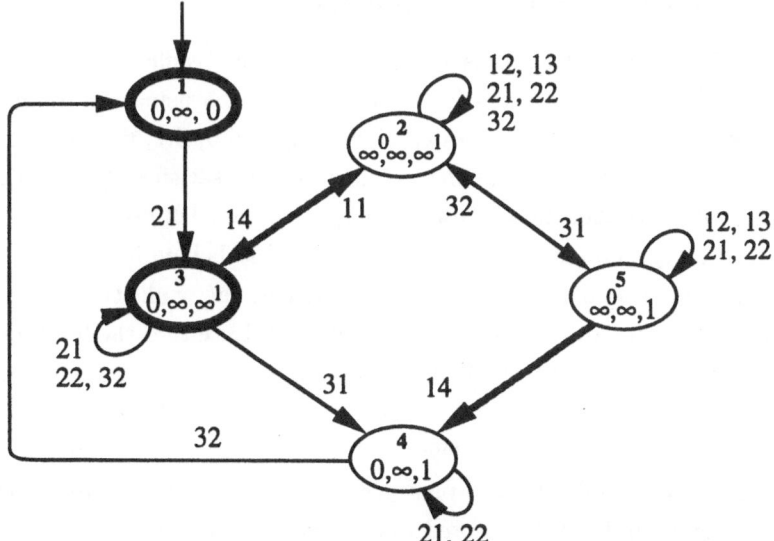

Figure 6.3: *a−machine* corresponding to dynamic specification D3.

closed. In these formulas, the "next transitions" not allowed to occur are specified by negation.

- Specification D4a. Avoid cyclic operation of valves when level measurement is not minimum.

  If DV is open, FV is closed and level measurement is not minimum, then the next transition must not be DV closing or FV opening. This will force the controller to wait for a change in the level before issuing any command to the system and thus avoiding cyclical operation of the valves. Note that if the level is empty, any of the prohibited transitions can occur.

  $$(\infty^0, 0, 1) \rightarrow \bigcirc[(\tau \neq 21) \wedge (\tau \neq 32)] \tag{6.7}$$

- Specification D4b. Avoid cyclic operation of valves when level measurement is empty.

  On the other hand, if FV is open, DV is closed and level measurement is empty, then FV must not close and DV must not open.

  $$(0, 1, 0) \rightarrow \bigcirc[(\tau \neq 22) \wedge (\tau \neq 31)] \tag{6.8}$$

The corresponding *a-machines* for these specifications are shown in figs 6.4 and 6.5 respectively.

6.2.3   Step 4.- Controller Synthesis

Given the five *a-machines* representing the dynamic specifications D1, D2, D3, D4a and D4b, a local controller is synthesised for each using the model of fig. 2.13. These local controllers are depicted in figs. 6.6, 6.7, 6.8, 6.9 and 6.10. Within each figure, the part of the *a-machine* that satisfies its specifications is framed within dashed lines. Note that specification D2 gives rise to a local controller that generates conditionally controllable behaviour in the system. When FV is open and normal level is detected in the tank, three transitions may occur according to the process model of fig. 2.13: either the level increases to maximum, FV is closed or DV is opened. Specification D2 prescribes only the closing of FV. Therefore, the existence of controllable behaviour from such a *state* is conditional upon the level measurement not indicating maximum. This is shown in fig. 6.7 by the presence of the state depicted by a gray line in which maximum level is detected and transition number 12 (level changing from normal to maximum) leading to it. The remaining specifications generate only controllable behaviour with respect to the process model.

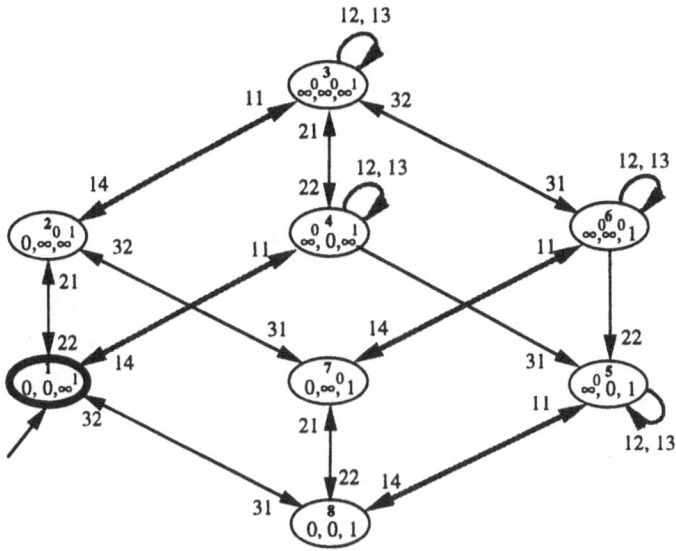

Figure 6.4: *a–machine* corresponding to dynamic specification D4a.

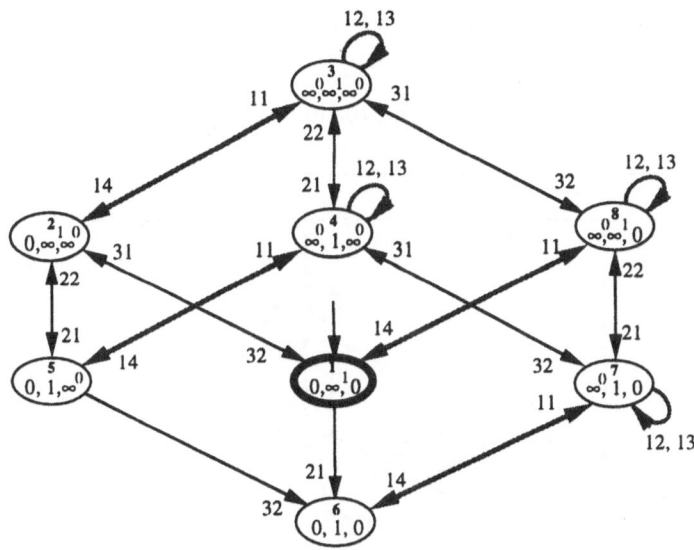

Figure 6.5: *a–machine* corresponding to dynamic specification D4b.

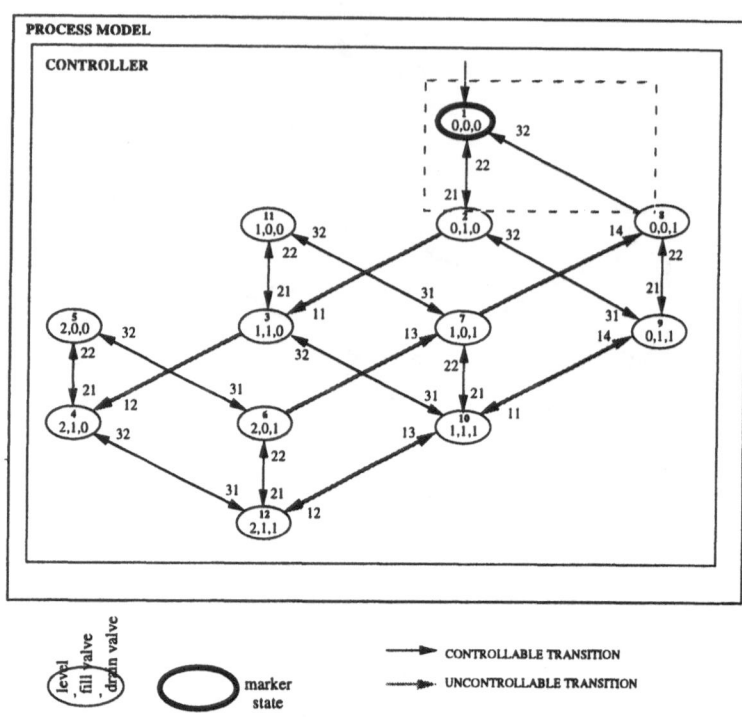

Figure 6.6: *a–machine* corresponding to the local controller for specification 1.

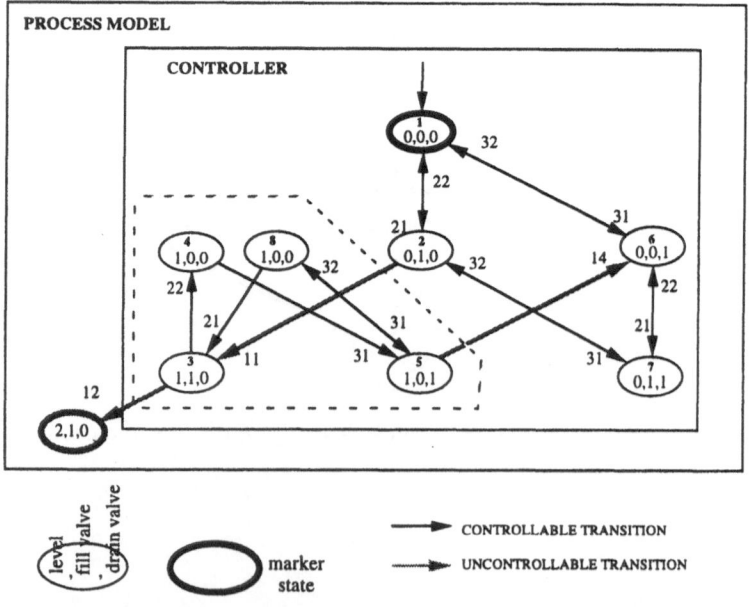

Figure 6.7: *a–machine* corresponding to the local controller for specification 2.

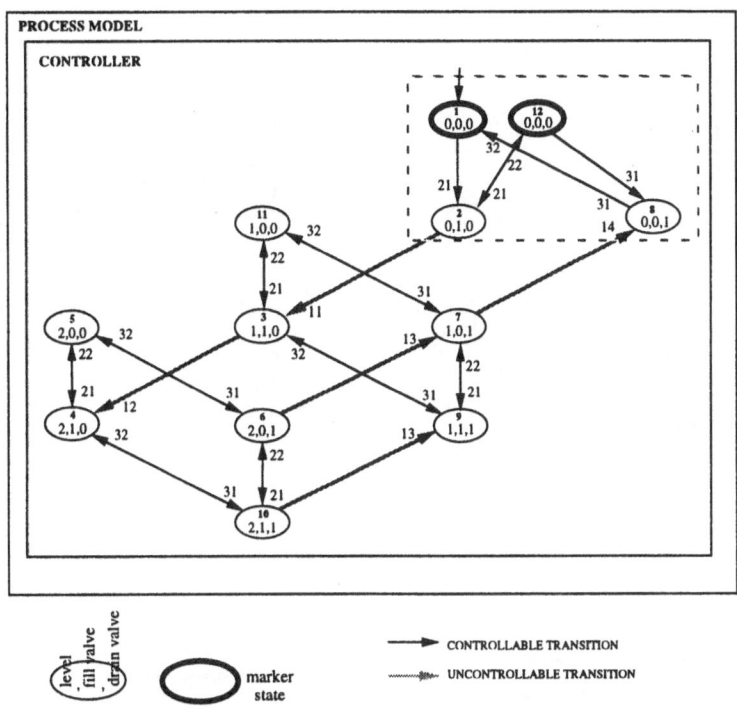

Figure 6.8: *a–machine* corresponding to the local controller for specification 3.

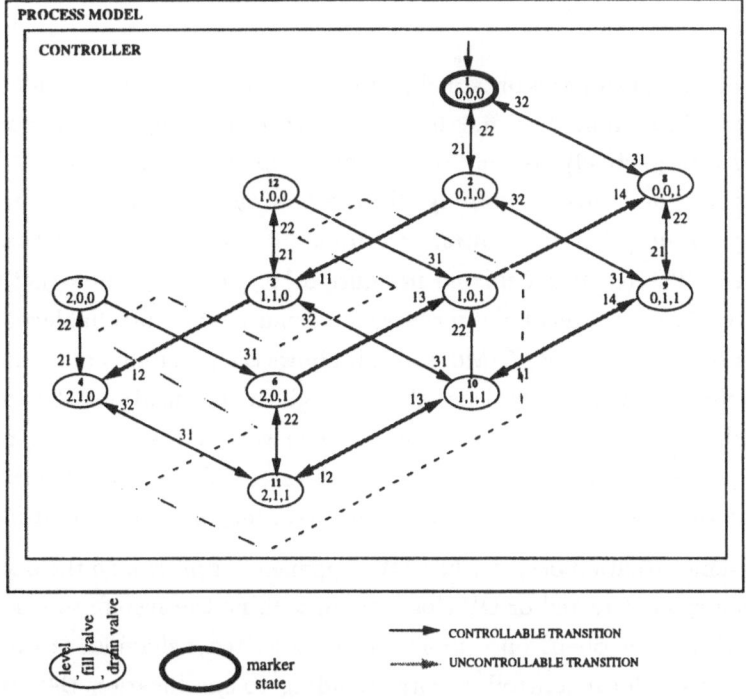

Figure 6.9: *a–machine* corresponding to the local controller for specification 4a.

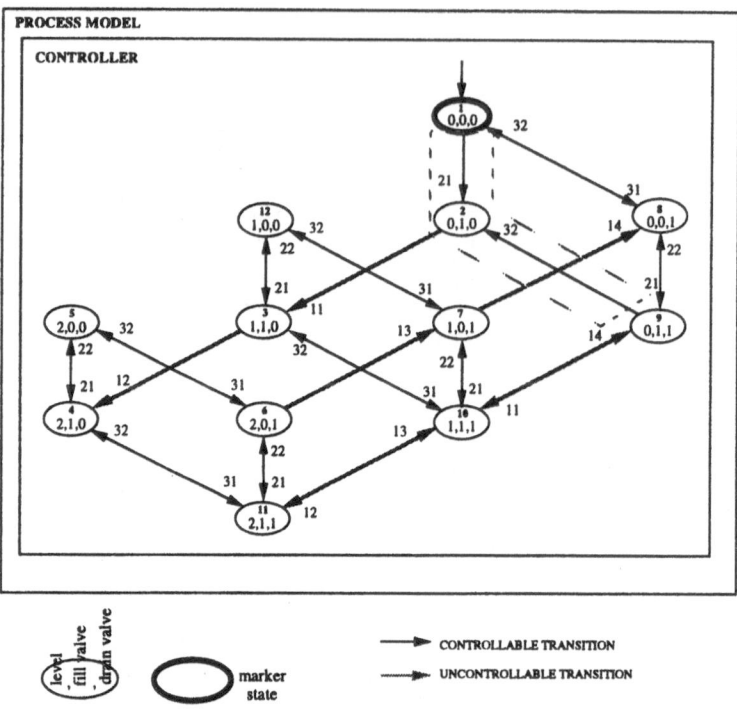

Figure 6.10: *a–machine* corresponding to the local controller for specification D4b.

The global controller is obtained by the *controller conjunction* of all the local controllers. Examining figs. 6.6, 6.7 and 6.8 corresponding to local controllers 1, 2 and 3 respectively, is easy to see that loops may exist where valves are being opened and closed infinitely. To see how these loops can be generated by the controller, first the *controller conjunction* is performed for these three controllers. The result is depicted in figure 6.11. The resultant structure does not satisfy the definition of *global controller* because two controllable transitions can be executed from *state* 7. Moreover, it allows the possibility of the indefinite valve cycling, although it satisfies the three given specifications. For instance, if normal level is not indicated in *state* 2, then valve FV may close (transition 22) driving the system to *state* 8 from where valve FV will open (transition 21) returning the system again to *state* 2. The cycle may be repeated *ad infinitum*.

The same situation occurs when DV is opened. From *state* 5 the decrease in the level may be detected or DV closes again sending the system to *state* 7 from where either valve opens once more. This undesired behaviour is avoided by introducing the local controllers corresponding to specification D4a and D4b.

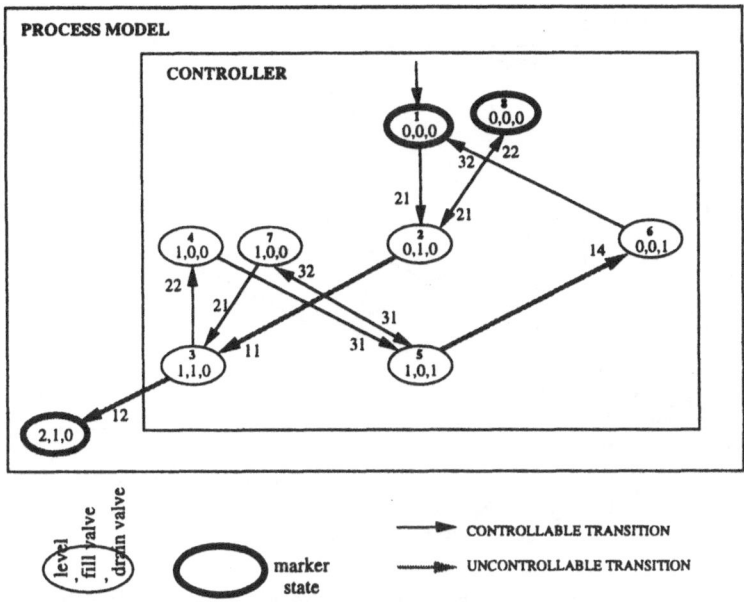

Figure 6.11: *a–machine* corresponding to the *controller composition* of local controllers 1, 2 and 3.

The result of the *controller conjunction* of these controllers and the resultant controller for specifications D1, D2 and D3 (fig. 6.11) is depicted in fig 6.12. As can be seen in this figure, the occurrence of events is strictly sequential complying with the definition of *global controller* and with the required operation of the buffer tank. The only *state* from where controllable behaviour is conditional to the no execution of an uncontrollable transition is *state* 3 from where the level measurement can reach its maximum value.

6.2.4   Controller Implementation and Simulation Results

6.2.4.1   Controller Implementation

The objective of the synthesis procedure is to find a *global controller* that maps each system *state* to at most one controllable transition. Assuming that the specification step was done successfully, if only one controllable transition (control command) is associated to a controller state, there is no choice and the controller must execute it. If one or several uncontrollable transitions exist in the same controller *state*, the controller must wait to receive from the measuring devices the transition to be executed reaching a new controller *state*. However, if one or several uncontrollable transitions and a controllable transition exist

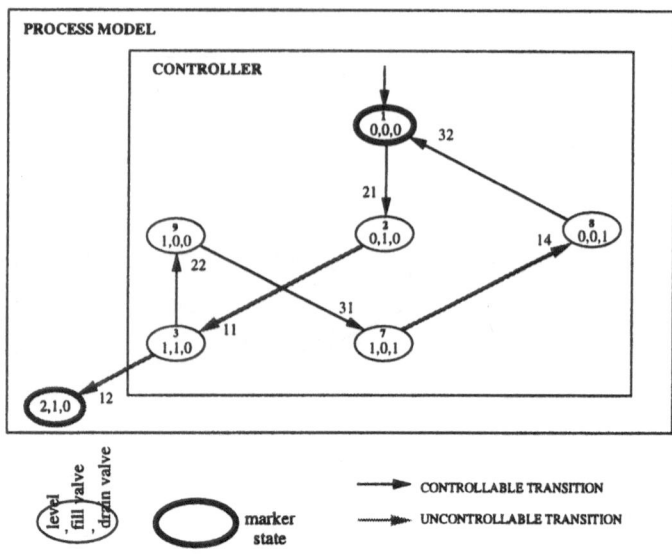

Figure 6.12: *a–machine* corresponding to the global controller for the buffer system.

in a controller *state*, it means that no specification prescribed the solely execution of the controllable transition from the given *state*. Moreover, the control model by itself does not have enough information to resolve this case. This indeterminacy is solved in the implementation of the controller by assigning priority of execution to the uncontrollable transitions and defining a period of time in which uncontrollable transitions can be executed. If none of the uncontrollable transitions are executed (i.e, the process does not generate any of the uncontrollable transitions as a response) after a certain period of time (identified as a sampling time), the controllable transition (control command) is then executed. Of course, other mechanisms to resolve the indeterminacy could in principle be envisaged. Having in mind the above considerations, the *a–machine* describing such a controller is implemented as a g*PROMS* model with the following components (see figs. 6.17, 6.18, 6.19 and 6.20):

- A **SELECTOR** variable **STATE** corresponding to the controller *state*. The values that the **SELECTOR** variable can take correspond to the *state* labels.

- A variable **STATE_NUMBER** to monitor and record the *state* in which the controller is.

- A variable flag DETECT to signal when a response from the system has been detected or a control command has been scheduled to the process from a given controller state.

Elementary g*PROMS* tasks are defined for each controllable transition of the elementary models such as the opening of a valve. Using these elementary tasks as building blocks, hierarchical g*PROMS* tasks, identified as "controller tasks", are constructed for each controller *state*, in which the detection of process responses (uncontrollable transitions) is performed by the change of suitable **SELECTOR** variables in the generic g*PROMS* models. The corresponding transition function to switch the **SELECTOR** variable STATE to a new controller *state* as a function of the executed task (transition) is also included. Two different types of controller tasks are identified:

- Tasks which are composed of only one controllable transition. In these tasks, the controllable transition must be immediately scheduled after reaching the associated controller *state*.

- Mixed tasks, which always contain uncontrollable transitions to be executed and may contain one controllable transition. In these tasks, the controller awaits for a process response that corresponds to one of the uncontrollable transitions associated to the controller *state*. If, after the sampling time has elapsed and no transition corresponding to those in the current controller *state* has been detected, the controllable transition is scheduled to the process. If no controllable transition is associated to the current controller *state*, the controller is sent to a fail *state* and the simulation is stopped.

Typical controller tasks are shown in figs. 6.17, 6.18, 6.19 and 6.20. The **SCHEDULE** section of each controller task commences with an **IF** instruction which checks if the **SELECTOR** variable STATE matches with the current controller *state* and if no process responses have been detected or control actions have been executed in the current sampling period. If the argument of the **IF** instruction is not satisfied the whole controller task is skipped. Else, the **THEN** argument of the task is executed. If the controller task is mixed (see for instance, task 2 in fig. 6.17), a set of **IF** conditions follows whose arguments are the uncontrollable transitions (i.e. expected process responses) in the current controller *state*. The **IF** conditions are checked sequentially. If one of the **IF** conditions is satisfied, the instructions within the **THEN** argument are executed. They perform the following actions

- Reset the sampling time to zero.

- Activate the variable flag **DETECT** to indicate the detection of a process response (uncontrollable transition) or the execution of a control command (controllable transition).

- Execute the transition function which switches the **SELECTOR** variable **STATE** to a new value.

If an elementary g*PROMS* task representing a controllable transition exists together with uncontrollable transitions in the current controller task and no uncontrollable transition is detected after reaching the maximum sampling time, then the elementary g*PROMS* task is executed. Finally, in order to avoid infinite waiting period for a transition to occur in mixed tasks, an **IF** instruction with a maximum waiting time, **MAX_WAIT_TIME**, is introduced at the end of the task. If the controller has been in the same state for longer than **MAX_WAIT_TIME** the system is sent to a "fail" state and the simulation is terminated. In the case of controllable tasks (see for instance, task 1 in fig. 6.17), the argument of the **THEN** section consists on an elementary task, the reset of the sampling time, the activation of the flag **DETECT** and the transition function. The **SCHEDULE** in the **PROCESS** section consists of the sequential execution of each of the controller tasks (see figs. 6.21 and 6.22).

6.2.4.2   g*PROMS* Model for the Buffer System

Generic models for the on/off valves and the surge tank were constructed and are shown in figs. 6.13, 6.14 and 6.15. The discrete states of these hybrid discrete/continuous models were defined so as to match the discrete–event models (*a–machines*) utilised for the controller synthesis. The *s_valve* model in fig. 6.13 describes the flow through the valve as a function of pressure drop and position of the valve while fig. 6.14 shows the elementary tasks that must be executed to open and close the valve. The *atm_tank* model in fig. 6.15 describes the change in the volume of the tank and the pressure in the outlet of the tank using mass and energy balances. Piping and discharge hydraulics are lumped within the discharge constants of the tank and valves.

The description of discrete states is effected using **SELECTOR** variables within the models. For instance, in the *atm_tank* model (fig. 6.15) the **SELECTOR** variable level_switch can take the values *EMPTY*, *NORMAL* and *HIGH*. In the case of the valves, the **SELECTOR** variable v_status can take the values *OPEN* or *CLOSED*. Tank and valve models are instantiated to appropriate components of the process forming the *tank* model which is shown

```
#
#============================================================
#
# Model of a solenoid valve
#
#============================================================
#
MODEL s'valve
PARAMETER
    VALVE'CONSTANT              AS REAL
VARIABLE
    position                   AS fraction
    flow                       AS flow'rate
    press'in                   AS pressure
    press'out                  AS pressure
    delta'p                    AS pressure
    c'action                   AS NoType
SELECTOR
    v'status                   AS (closed, open) DEFAULT closed
STREAM
    input  : flow, press'in    AS Mainstream
    output : flow, press'out   AS Mainstream
SET
    VALVE'CONSTANT:= 40 ;

EQUATION
# Flowrate / Delta P relationship
    flow = position*VALVE'CONSTANT*SGN(delta'p)*SQRT(ABS(delta'p)) ;
# Delta'P definition
    delta'p = press'in - press'out ;
    CASE v'status OF
        WHEN closed      : position = 0.0;
                     SWITCH TO open      IF c'action = 1 ;
        WHEN open        : position = 1.0;
                     SWITCH TO closed    IF c'action = 0 ;
    END # case
END # model s'valve
```

Figure 6.13: *gPROMS* generic model of an on/off valve

in fig. 6.16. It is composed of a surge tank, the on/off feed valve *FV* and the on/off exit valve *EV*.

The controller shown in fig. 6.12 is automatically translated into g*PROMS* code. It results in 6 g*PROMS* controller tasks, one for each controller *state*. The controller model and associated controller tasks are shown in figs. 6.17, 6.18, 6.19 and 6.20. The **PROCESS** section and its **SCHEDULE** are shown in figs. 6.21 and 6.22. Following the initialisation sections, the schedule starts with a **WHILE** instruction to execute the operating cycle several times (for demonstration purposes). The first transition to be executed is the opening of FV (control task 1 in fig. 6.17). Then the controller waits for the process response indicating that the tank is at its normal level (control task 2 in fig. 6.18). If such a signal is not detected within the maximum sampling time, the operation is terminated. If the signal is detected (**IF (tank.surge_tank.level_switch = tank.surge_tank.normal) THEN SEQUENCE**) the procedure continues clos-

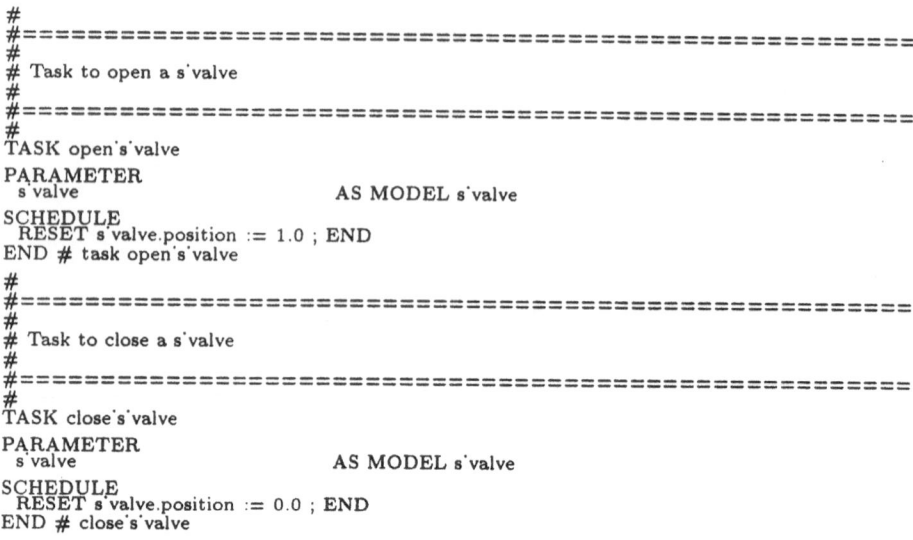

```
#
#=============================================================
#
# Task to open a s'valve
#
#=============================================================
#
TASK open's'valve
PARAMETER
   s'valve                          AS MODEL s'valve
SCHEDULE
   RESET s'valve.position := 1.0 ; END
END # task open's'valve
#
#=============================================================
#
# Task to close a s'valve
#
#=============================================================
#
TASK close's'valve
PARAMETER
   s'valve                          AS MODEL s'valve
SCHEDULE
   RESET s'valve.position := 0.0 ; END
END # close's'valve
```

Figure 6.14: *gPROMS* tasks to open and close an on/off valve

ing FV (`close_s_valve(valve IS tank.fill_valve)`) (control task 3 in fig. 6.18) and opening DV (control task 4 in fig. 6.19) to commence the draining of the tank. The schedule awaits the *EMPTY* signal from the level switch (control task 5 in fig. 6.19) and closes DV (control task 6 in fig. 6.19) to finish the drain process. The operating sequence then repeats for 2000 time units. Figure 6.23 shows the tank level trajectory followed by the continuous dynamic system when controlled by the procedural controller.

6.3   Metering Tank System

The second example is an evolution of the first to a more realistic case. The buffer tank with an extra level switch is used to measure a fixed amount of liquid to be dispensed downstream (fig. 6.24). Valves FV and DV may fail while opening or closing. An extra valve, RV, is supplied to drain the excess of liquid in the tank in case of valve failures. A controller is required to guarantee that under any circumstance the proper amount of liquid will be dispensed downstream. Valves FV and DV are modelled with increased detail. Uncontrollable and controllable transitions are permitted to exist within the same elementary model, as opposed to all previous examples in which different elementary models were the source of either controllable or uncontrollable transitions. Therefore, the controller will have some *states* with mixed controllable

```
#
#===============================================================
#
# Model of atmospheric tank
#
#===============================================================
#
MODEL atm'tank
PARAMETER
    ATM'PRESS                    AS REAL
    AREA                         AS REAL
    DENSITY                      AS REAL
    LEVEL'TOL                    AS REAL
    MAXVOL                       AS REAL
    NORMALVOL                    AS REAL
    HIGHVOL                      AS REAL
    EMPTYVOL                     AS REAL

VARIABLE

    flow'in, flow'out            AS flow'rate
    volume                       AS volume
    over'flow                    AS flow'rate
    liquid'height                AS length
    press'in                     AS pressure
    press'out                    AS pressure

STREAM
    input      : flow'in, press'in     AS Mainstream
    output     : flow'out, press'out   AS Mainstream

SELECTOR
    level'switch                 AS (empty, normal, high)

EQUATION

    $volume = (flow'in - (flow'out+over'flow))/100.0 ;
    liquid'height = volume/AREA;

# Hydrostatic pressure
    press'out = ATM'PRESS + 9.81*DENSITY*liquid'height/1E5 ;

# Determining the value of level'switch
    CASE level'switch OF
       WHEN high    : over'flow = 0 ;
                SWITCH TO normal  IF volume LT  (NORMALVOL + LEVEL'TOL) ;
       WHEN normal : over'flow = 0 ;
                SWITCH TO high     IF volume GE (HIGHVOL -LEVEL'TOL) ;
                SWITCH TO empty   IF volume LE EMPTYVOL ;
       WHEN empty  : over'flow = 0 ;
                SWITCH TO normal  IF (volume GE NORMALVOL) and
                                  (volume LT (NORMALVOL+ LEVEL'TOL));
    END # case
END # model atm'tank
```

Figure 6.15: *gPROMS* model of an atmospheric tank

and uncontrollable transitions. Furthermore, apart from normal operation, abnormal behaviour is also considered. Marked *states* in the process model are defined as a function of the marked *states* of the elementary process models. A *state* in the process model is marked if all the *states* in the elementary models from which the *state–variable* values are taken to form the current *state*, are marked. In order to simplify the handling of marked *states* in the specifications, all the *states* in the corresponding dynamic specification *a–machines* are assumed to be marked.

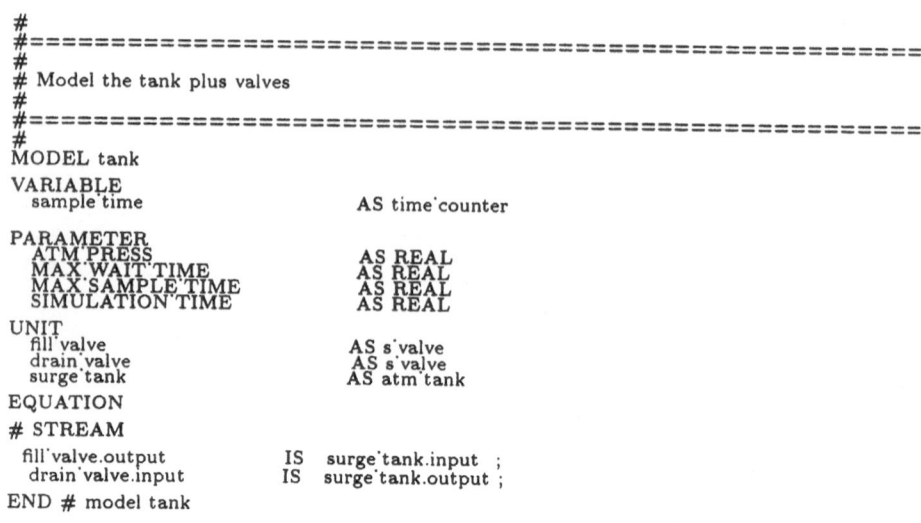

```
#
#==================================================================
#
# Model the tank plus valves
#
#==================================================================
#
MODEL tank
VARIABLE
   sample time                    AS time counter

PARAMETER
   ATM PRESS                      AS REAL
   MAX WAIT TIME                  AS REAL
   MAX SAMPLE TIME                AS REAL
   SIMULATION TIME                AS REAL
UNIT
   fill valve                     AS s valve
   drain valve                    AS s valve
   surge tank                     AS atm tank
EQUATION
# STREAM

   fill valve.output       IS    surge tank.input  ;
   drain valve.input       IS    surge tank.output ;
END # model tank
```

Figure 6.16: *gPROMS* model of the metering system

6.3.1 Process Description

The metering system is shown in fig. 6.24. It consists of a tank with a level
measuring device with four level switches to indicate when the tank is empty,
at the proper discharge level (normal level), above the discharge level (high
level) or at its maximum. Liquid is fed into the tank using the on/off valve
FV and discharged through on/off valve EV. These two valves are prone to fail
while either opening or closing. Each valve is also able to send responses to the
controller indicating if it has opened or closed successfully or it is stuck closed
or open. Also, once the valve has been repaired after being stuck, it emits a
signal indicating its availability. To cope with the eventual failure of any of
these two valves and to adjust the liquid to the proper discharge level if required,
the on/off valve RV is provided. Valve RV has been designed to handle the
maximum expected upstream flowrate when the tank is at its maximum level in
order to avoid spilling. It is connected to a dump tank where the liquid can be
collected for reuse. To reduce the size of the model and make it more tractable,
it is assumed that this valve does not fail and only takes two possible states,
open or closed. An on/off switch is provided to start and stop the process.
The initial and final states of the operation are when all the valves are closed,
the tank is empty and the on/off switch is in the off position. To start the
operation, the switch takes the on position and to terminate, it changes to off.
Once the switch is off, the process must reach a safe state (i.e. level empty and

all valves closed). The objective of the synthesis is to find a controller to drive
the operation of the metering tank so as to guarantee that the volume being
dispensed to the downstream process is always properly measured (i.e. the
level measurement must indicate normal level before starting the discharge). If
either FV or EV fails, the controller must signal for the repair crew to service
the faulty valve and await acknowledgement that the repair has been effected.
If FV fails to close, the controller must use RV to avoid spillage while FV
is being repaired. Also, RV must be used to remove from the metering tank
any excess of liquid. The controller is synthesised using the four steps of the
synthesis procedure proposed in chapter 5.

```
#
#===========================================================
#
# Procedural Controller Model
#
#===========================================================
#
MODEL fsa
VARIABLE
   state'number              AS NoType
   detect                    AS NoType
SELECTOR
   state                     AS ( s1, s2, s3, s4, s5, s6, s7)
EQUATION
   CASE state OF
      WHEN s1 : state'number = 1;
      WHEN s2 : state'number = 2;
      WHEN s3 : state'number = 3;
      WHEN s4 : state'number = 4;
      WHEN s5 : state'number = 5;
      WHEN s6 : state'number = 6;
      WHEN s7 : state'number = 7;
   END # case
END # model fsa
#
#===========================================================
#
# Task state 1
#
#===========================================================
#
TASK Check'state'1
PARAMETER
   fsa          AS MODEL fsa
   tank         AS MODEL tank
SCHEDULE
   IF (fsa.state = fsa.s1) and (fsa.detect = 0) THEN
      SEQUENCE
         open's'valve(s'valve IS tank.fill'valve);
         RESET fsa.detect := 1 ; END
         SWITCH fsa.state := fsa.s2; END
      END # sequence
   END # if
END # task
```

Figure 6.17: *gPROMS* model of the controller (part A).

```
#
#============================================================
#
# Task state 2
#
#============================================================
#
TASK Check`state`2
PARAMETER
  fsa           AS MODEL fsa
  tank          AS MODEL tank
SCHEDULE
  IF (fsa.state = fsa.s2) and (fsa.detect = 0) THEN
     SEQUENCE
        IF (tank.surge`tank.level`switch=tank.surge`tank.normal) and
           (fsa.detect = 0) THEN
           SEQUENCE
              RESET tank.sample`time := OLD(TIME) ; END
              RESET fsa.detect := 1 ; END
              SWITCH fsa.state := fsa.s3; END
           END # sequence
        END # if
        IF (TIME - tank.sample`time GT tank.MAX`SAMPLE`TIME) and
           (fsa.detect = 0) THEN
           SEQUENCE
              RESET tank.sample`time := OLD(TIME) ; END
              RESET fsa.detect := 1 ; END
              SWITCH fsa.state := fsa.s7 ; END
           END # parallel
        END # if
     END # sequence
  END # if
END # task
#
#============================================================
#
# Task state 3
#
#============================================================
#
TASK Check`state`3
PARAMETER
  fsa           AS MODEL fsa
  tank          AS MODEL tank
SCHEDULE
  IF (fsa.state = fsa.s3) and (fsa.detect = 0) THEN
     SEQUENCE
        close`s`valve(s`valve IS tank.fill`valve);
        RESET fsa.detect := 1 ; END
        SWITCH fsa.state := fsa.s4; END
     END # sequence
  END # if
END # task
```

Figure 6.18: *gPROMS* model of the controller (part B).

```
#
#==========================================================
#
# Task state 4
#
#==========================================================
#
TASK Check'state'4
PARAMETER
   fsa            AS MODEL fsa
   tank           AS MODEL tank
SCHEDULE
   IF (fsa.state = fsa.s4) and (fsa.detect = 0) THEN
     SEQUENCE
        open's'valve(s'valve IS tank.drain'valve);
        RESET fsa.detect := 1 ; END
        SWITCH fsa.state := fsa.s5; END
     END # sequence
   END # if
END # task
#
#==========================================================
#
# Task state 5
#
#==========================================================
#
TASK Check'state'5
PARAMETER
   fsa            AS MODEL fsa
   tank           AS MODEL tank
SCHEDULE
   IF (fsa.state = fsa.s5) and (fsa.detect = 0) THEN
     SEQUENCE
        IF (tank.surge'tank.level'switch = tank.surge'tank.empty) and
           (fsa.detect = 0) THEN
          SEQUENCE
             RESET tank.sample'time := OLD(TIME) ; END
             RESET fsa.detect := 1 ; END
             SWITCH fsa.state := fsa.s6; END
          END # sequence
        END # if
        IF (TIME - tank.sample'time GT tank.MAX'SAMPLE'TIME) and
           (fsa.detect = 0) THEN
          SEQUENCE
             RESET tank.sample'time := OLD(TIME) ; END
             RESET fsa.detect := 1 ; END
             SWITCH fsa.state := fsa.s7 ; END
          END # parallel
        END # if
     END # sequence
   END # if
END # task
#
#==========================================================
#
# Task state 6
#
#==========================================================
#
TASK Check'state'6
PARAMETER
   fsa            AS MODEL fsa
   tank           AS MODEL tank
SCHEDULE
   IF (fsa.state = fsa.s6) and (fsa.detect = 0) THEN
     SEQUENCE
        close's'valve(s'valve IS tank.drain'valve);
        RESET fsa.detect := 1 ; END
        SWITCH fsa.state := fsa.s1; END
     END # sequence
   END # if
END # task
```

Figure 6.19: *gPROMS* model of the controller (part C).

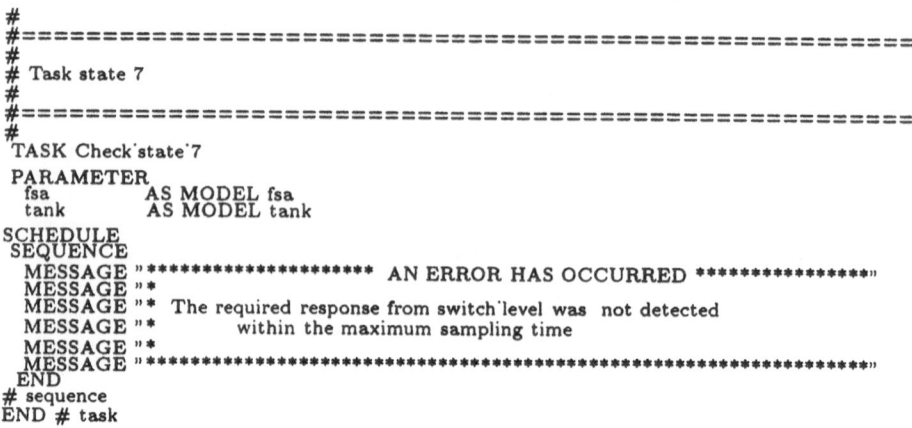

```
#
#===========================================================
#
# Task state 7
#
#===========================================================
#
 TASK Check state 7
 PARAMETER
   fsa          AS MODEL fsa
   tank         AS MODEL tank
 SCHEDULE
  SEQUENCE
   MESSAGE " *******************  AN ERROR HAS OCCURRED  ***************"
   MESSAGE " *
   MESSAGE " *  The required response from switch level was  not detected
   MESSAGE " *          within the maximum sampling time
   MESSAGE " *
   MESSAGE " ****************************************************************"
  END
 # sequence
 END # task
```

Figure 6.20: *gPROMS* model of the controller (part D).

6.3.2   Step 1.- Process Modelling

The process is divided into five elementary process components: level measurement, valve FV, valve EV, valve RV and on/off switch. The *a–machines* for each component are described in table 6.3.2. The level measurement is modelled as the *a–machine* M_1 shown in fig. 6.25. Its *state–variable level_status* takes four different values, giving rise to four *states*, each corresponding to a level switch in the tank. *States* 1 and 2 in which *state–variable level_status* takes the value of empty or normal are marked. The transitions connecting the *states* are uncontrollable. Valves FV and RV are modelled as *a–machines* M_2 and M_3 respectively with seven states as illustrated in fig. 6.26 in which the *state–variable valve_status* can take the following values: valve open or closed, in the process of opening or closing, stuck opened or closed and under repair. If the valve is open, a control command is issued to force the valve to close. Afterwards, the valve may uncontrollably either fail to open or fulfill the control instruction. If it fails and becomes stuck open, the next control command will drive the valve to the repair state but with the valve still open. Following repair, the valve returns to the closed state. The same behaviour occurs when the valve is closed. If the valve fails to execute the opening action, it will reach a state in which it is stuck closed and a control command must be issued to repair it, eventually reaching the closed state again. Note that control actions (controllable transitions) are modelled together with the responses of the elementary component (the uncontrollable transitions). Valve RV is modelled by *a–machine* M_4 shown in fig. 6.27. It can only be open

```
#
#=========================================================
#
# Simulation. Cyclic operation.
#
#=========================================================
#
PROCESS CYCLE
UNIT
    tank'unit              AS tank
    controller             AS fsa
SET
    WITHIN tank'unit DO
        ATM'PRESS              := 1.01325 ;
        MAX'SAMPLE'TIME        := 207 ;
        MAX'WAIT'TIME          := 203 ;
        SIMULATION'TIME        := 200 ;
        WITHIN surge'tank DO
            DENSITY                := 1000;
            AREA                   := 1;
            LEVEL'TOL              := 0.5;
            MAXVOL                 := 30;
            NORMALVOL              := 15;
            HIGHVOL                := 20;
            EMPTYVOL               := 1;
        END # within
    END # within
ASSIGN
    WITHIN tank'unit DO
        WITHIN fill'valve DO
            c'action               := 0;
            press'in               := 5.0 ;
            press'out              := 1.101325 ;
        END # fill'valve
        WITHIN surge'tank DO
            sample'time            := 0.0;
        END # surge'tank
        WITHIN drain'valve DO
            c'action               := 0;
            press'out              := 1.101325 ;
        END # drain'valve
    END # within
    WITHIN controller DO
        detect                 := 0 ;
    END # controller
```

Figure 6.21: *gPROMS* process and schedule of the buffer system (part A).

or closed and it changes its status in response to the controllable transitions (control commands) "open valve" or "close valve". The only marked *state* in each of the three valves is when it is closed. In this way, only circumstances in which there is no change in the liquid level are considered as possible *states* in which the operation can be terminated or held. The on/off switch is modelled as the two–*states a–machine* M_5 shown in fig. 6.28. Both states are marked. Transitions are uncontrollable to consider the change in the switch as an non-deterministic external action (e.g., an operator changes the switch position to off unexpectedly due to problems in the process upstream).

Given that the set of transitions of the five elementary models are disjoint, a model considering all the possible behaviour is obtained by the *asynchronous product* of the five elementary models following the same procedure as in the

```
PRESET
  WITHIN tank unit DO
    sample time        := 0;
    WITHIN fill valve DO
      flow             := 0;
    END # fill valve
    WITHIN surge tank DO
      liquid height    := 0.1;
      $VOLUME          := 0.1 ;
    END # surge tank
    WITHIN drain valve DO
      flow             := 0;
    END # drain valve
  END # within
SELECTOR
  WITHIN tank unit DO
    WITHIN surge tank DO
      level switch     := empty ;
    END # surge tank
  END # within
  WITHIN controller DO
    state              := s1 ;
  END # controller
INITIAL
  WITHIN tank unit DO
    WITHIN surge tank DO
      volume           = 0 ;
    END # surge tank
  END # within
SCHEDULE
  WHILE TIME ¡ tank unit.SIMULATION TIME DO
    SEQUENCE
      Check state 1(tank IS tank unit, fsa is controller);
      Check state 2(tank IS tank unit, fsa is controller);
      Check state 3(tank IS tank unit, fsa is controller);
      Check state 4(tank IS tank unit, fsa is controller);
      Check state 5(tank IS tank unit, fsa is controller);
      Check state 6(tank IS tank unit, fsa is controller);
      RESET controller.detect := 0 ;END
      CONTINUE FOR 1
    END # sequence
  END # while
END # schedule
```

Figure 6.22: *gPROMS* process and schedule of the buffer system (part B).

past three examples. The obtained structure contains 784 *states*. The order of the elementary *state–variables* in each *state* is: (level status, FV status, EV status, RV status, switch status).

As in the previous example, causal behaviour in the level is introduced by means of TL formulas in order to consider only physically possible behaviour and to reduce the number of transitions from each *state*. The size of the discrete–event model makes it difficult to precisely determine conditions for the change of level measurement and only two scenarios are considered:

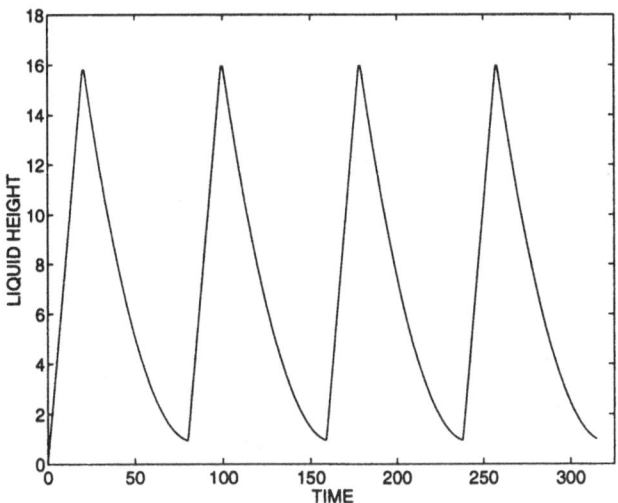

Figure 6.23: Level trajectory in buffer system.

- When FV is closed or being opened, the level cannot increase. This is
  stated by the following TL formula

$$(\infty_1, \infty_2^{2,3,4,5,6}, \infty_3, \infty_4, \infty_5) \rightarrow \bigcirc[(\tau \neq 11) \wedge (\tau \neq 12) \wedge (\tau \neq 13)] \quad (6.9)$$

- When EV and RV are closed or EV is being opened and RV is closed,
  the level cannot decrease. The following TL formula represents such
  behaviour

$$(\infty_1, \infty_2, \infty_3^{2,3,4,5,6}, 0, \infty_5) \rightarrow \bigcirc[(\tau \neq 14) \wedge (\tau \neq 15) \wedge (\tau \neq 16)] \quad (6.10)$$

Note that the LHS state of each TL formula has two complement states
giving rise to the following formulas:

$$(\infty_1, 0, \infty_3, \infty_4, \infty_5) \rightarrow \bigcirc[(\tau \neq 11) \wedge (\tau \neq 12) \wedge (\tau \neq 13)] \qquad (6.11.a)$$

$$(\infty_1, 1, \infty_3, \infty_4, \infty_5) \rightarrow \bigcirc[(\tau \neq 11) \wedge (\tau \neq 12) \wedge (\tau \neq 13)] \qquad (6.11.b)$$

$$(\infty_1, \infty_2, 0, 0, \infty_5) \rightarrow \bigcirc[(\tau \neq 14) \wedge (\tau \neq 15) \wedge (\tau \neq 16)] \qquad (6.12.a)$$

$$(\infty_1, \infty_2, 1, 0, \infty_5) \rightarrow \bigcirc[(\tau \neq 14) \wedge (\tau \neq 15) \wedge (\tau \neq 16)] \qquad (6.12.b)$$

These TL formulas are translated into their respective *a–machines*. The
process model is obtained by the *synchronous product* of these *a–machines* and
the result from the *asynchronous product* of the elementary components.

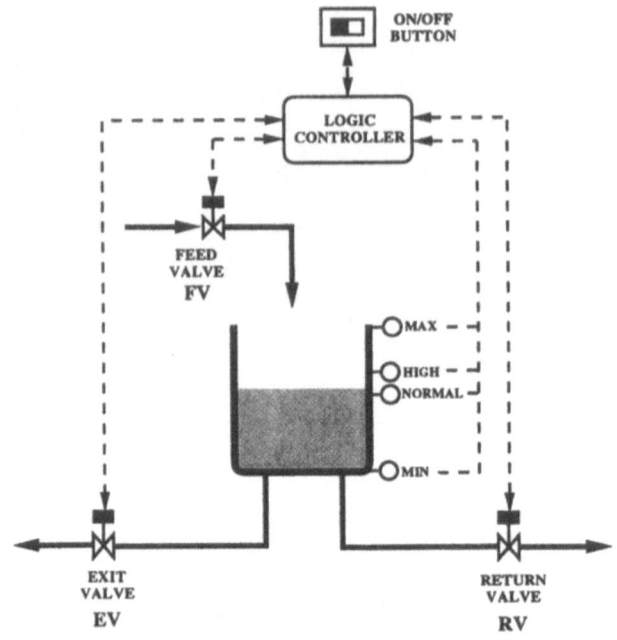

Figure 6.24: The metering system.

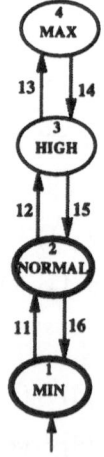

Figure 6.25: *a-machine* M_1 corresponding to the level measuring.

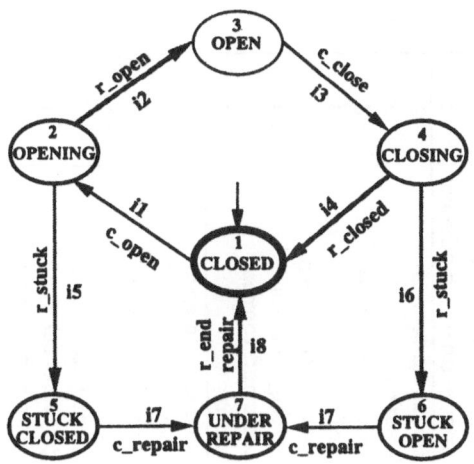

Figure 6.26: *a-machine* $M_i, i = 2, 3$, corresponding to valves FV and EV.

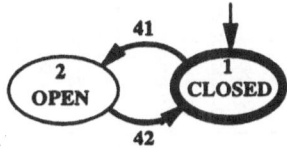

Figure 6.27: *a-machine* M_4 corresponding to valve RV.

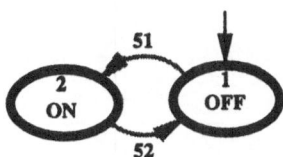

Figure 6.28: *a-machine* M_5 corresponding to on/off switch.

elementary component	a-machine	fig.	state-variable description	value	state label	transition label	transition description	from st	to st
level switch	M_1	6.25	level status	0: minimum	1	11*	from minimum to normal	1	2
				1: normal	2	12*	from normal to high	2	3
				2: high	3	13*	from high to maximum	3	4
				3: maximum	4	14*	from maximum to high	4	3
						15*	from high to normal	3	2
						16*	from normal to minimum	2	1
fill valve FV	M_2	6.26	valve status	0: closed	1	21	command to open	1	2
				1: being opened	2	22*	opening	2	3
				2: open	3	23	command to close	3	4
				3: being closed	4	24*	closing	4	1
				4: stuck open	5	25*	getting stuck open	2	5
				5: stuck closed	6	26*	getting stuck closed	4	6
				6: under repair	7	27	command to repair	5, 6	7
						28*	end repair	7	1
exit valve EV	M_3	6.26	valve status	0: closed	1	31	command to open	1	2
				1: being opened	2	32*	opening	2	3
				2: open	3	33	command to close	3	4
				3: being closed	4	34*	closing	4	1
				4: stuck open	5	35*	getting stuck open	2	5
				5: stuck closed	6	36*	getting stuck closed	4	6
				6: under repair	7	37	command to repair	5, 6	7
						38*	end repair	7	1
discharge valve RV	M_4	6.27	valve status	0: closed	1	41	command to open	1	2
				1: open	2	42	command to close	2	1
on/off switch	M_5	6.28	switch status	0: off	1	51*	command to switch on	1	2
				1: on	2	52*	command to switch off	2	1

Table 6.1: Elementary *a-machines* of the metering system (* = uncontrollable transition).

6.3.3   Step 2.- Specification Modelling

In the following paragraphs, the desired process behaviour described in subsection 6.3.1 is formalised and translated into forbidden states and TL formulas.

6.3.3.1   Static Specifications

In order to maintain a safe operation, if the level measurement is beyond its normal value, FV and EV must be guaranteed never to open or to be open. If FV is open, the level may increase unnecessarily while if EV is open, a wrong amount of liquid may be dispensed downstream. If the level measurement is higher than normal, the only valve that must be used is RV. Note that RV does not have to be used either if the tank is empty. This can be expressed in the following PL statements:

- If the level measurement is high or maximum, FV must never be open in order to avoid increasing the liquid in the tank unnecessarily.

$$q : (\infty_1^{0,1}, \infty_2^{0,3,4,5,6}, \infty_3, \infty_4, \infty_5) = \text{FALSE} \tag{6.13}$$

- If the level is high or maximum, EV must never be open in order to avoid discharge of the improper metered volume downstream.

$$q : (\infty_1^{0,1}, \infty_2, \infty_3^{0,3,4,5,6}, \infty_4, \infty_5) = \text{FALSE} \tag{6.14}$$

- If the level is minimum, RV must never be open.

$$q : (0, \infty_2, \infty_3, 1, \infty_5) = \text{FALSE} \tag{6.15}$$

In order to simplify the operation of the metering system, none of the valves must be allowed to operate simultaneously. That is, they must never change position simultaneously. Also, FV and EV must never be open at the same time. If one of the valves gets stuck, the other must not change position until the failed valve is repaired.

- Only one valve is allowed to be open at one time.

$$q : (\infty_1, 2, 2, \infty_4, \infty_5) = \text{FALSE} \tag{6.16}$$

$$q : (\infty_1, 2, 1, \infty_4, \infty_5) = \text{FALSE} \tag{6.17}$$

$$q : (\infty_1, 1, 2, \infty_4, \infty_5) = \text{FALSE} \tag{6.18}$$

$$q : (\infty_1, 2, \infty_3, 1, \infty_5) = \text{FALSE} \tag{6.19}$$

$$q : (\infty_1, \infty_2, 2, 1, \infty_5) = \text{FALSE} \tag{6.20}$$

- Valves FV and EV must not change position simultaneously.

$$q : (\infty_1, 1, 3, \infty_4, \infty_5) = \text{FALSE} \tag{6.21}$$

$$q : (\infty_1, 3, 1, \infty_4, \infty_5) = \text{FALSE} \tag{6.22}$$

- If FV fails, do not operate EV until FV is repaired. If FV fails or is under repair, it is possible that the level will be varying. Therefore, EV must not be allowed to open until FV is repaired.

$$q : (\infty_1, \infty_2^{0,1,2,3}, 1, \infty_4, \infty_5) = \text{FALSE} \tag{6.23}$$

$$q : (\infty_1, \infty_2^{0,1,2,3,6}, 3, \infty_4, \infty_5) = \text{FALSE} \tag{6.24}$$

- If EV fails, do not attempt to open FV until EV is repaired.

$$q : (\infty_1, 1, \infty_3^{0,1,2,3,6}, \infty_4, \infty_5) = \text{FALSE} \tag{6.25}$$

6.3.3.2 Dynamic Specifications

- **Normal Operation**

  - N1.- To start operation, the on/off switch must be turned on.

    $$(0,0,0,0,0) \rightarrow \bigcirc(\tau = 51) \tag{6.26}$$

  - N2.- Start filling the tank.
    If the level measurement is at minimum, the valves are closed and on/off switch is on, then the command to open FV must be issued.

    $$(0,0,0,0,1) \rightarrow \bigcirc(\tau = 21) \tag{6.27}$$

  - N3.- Changing from filling to emptying.

    * N3a.- If the level measurement is at minimum, FV is open, EV is closed and a change to normal level is detected, then the command to close FV must be issued.

    $$(0,2,0,\infty_4,\infty_5) \wedge (\tau = 11) \rightarrow \bigcirc(\tau = 23) \tag{6.28}$$

* N3b.- Once the tank is at its normal level and FV is closed, then the command to open EV must be issued.

$$(1, 0, 0, 0, \infty_5) \rightarrow \bigcirc(\tau = 31) \tag{6.29}$$

- N4.- Changing from emptying to filling.

If the level measurement is normal, FV is closed, EV is open, RV is closed and the minimum level value is detected in the tank, then the command to close EV must be issued.

$$(1, 0, 2, 0, \infty_5) \wedge (\tau = 16) \rightarrow \bigcirc(\tau = 33) \tag{6.30}$$

- N5.- Finishing operation.

    * N5a.- If on/off switch is turned off, eventually the system must reach the final state.

$$(\infty_1, \infty_2, \infty_3, \infty_4, 1) \wedge (\tau = 52) \rightarrow \Diamond(0, 0, 0, 0, 0) \tag{6.31}$$

    * N5b.- If on/off switch is turned off, then FV must never open again.

$$(\infty_1, \infty_2, \infty_3, \infty_4, 1) \wedge (\tau = 52) \rightarrow \Box(\tau \neq 21) \tag{6.32}$$

Note that the behaviour prescribed by this formula will not conflict with the starting of the tank filling prescribed in formula 6.27 because the only refined *state-variable* is the on/off switch while the rest are covered.

• **Abnormal Operation.**

    - A6.- Handling High and Maximum Levels.

        * A6a.- If RV is closed and a high level is detected, then open RV.

$$[(1, \infty_2, \infty_3, 0, \infty_5) \wedge (\tau = 12)] \rightarrow \bigcirc(\tau = 41) \tag{6.33}$$

        * A6b.- If DV is open and level goes back to normal, then close RV.

$$[(2, \infty_2, \infty_3, 1, \infty_5) \wedge (\tau = 15)] \rightarrow \bigcirc(\tau = 42) \tag{6.34}$$

    - A7.- Handling FV stuck.

    FV can get stuck either open or closed. Once the controller detects the valve stuck, it must immediately issue a repair command. This is prescribed by the following formula

$$(\infty_1, \infty_2^{0,1,2,3,6}, \infty_3, \infty_4, \infty_5) \rightarrow \bigcirc(\tau = 27) \tag{6.35}$$

This formula is decomposed into two formulas prescribing separately when FV is stuck open or stuck closed and the issuing of the repair command.

* A7a.- For the case when FV is stuck open, the following formula describes the issuing of the repair command.

$$(\infty_1, 4, \infty_3, \infty_4, \infty_5) \to \bigcirc(\tau = 27) \tag{6.36}$$

* A7b.- When FV is stuck closed, the following formula describes the issuing of the repair command.

$$(\infty_1, 5, \infty_3, \infty_4, \infty_5) \to (\bigcirc\tau = 27) \tag{6.37}$$

– A8.- Handling EV stuck.

Valve EV getting stuck is treated in the same way as dynamic specification A7.

$$(\infty_1, \infty_2, \infty_3^{0,1,2,3,6}, 2, \infty_5) \to \bigcirc(\tau = 37) \tag{6.38}$$

Again, this formula is decomposed in two formulas

* A8a.- Valve FV is stuck open.

$$(\infty_1, \infty_2, 4, \infty_4, \infty_5) \to \bigcirc(\tau = 37) \tag{6.39}$$

* A8b.- Valve FV is stuck open.

$$(\infty_1, \infty_2, 5, \infty_4, \infty_5) \to (\bigcirc\tau = 37) \tag{6.40}$$

6.3.4   Step 3.- Supervisory Structure Synthesis

The 13 PL statements representing forbidden states are eliminated from the process model. The pruned *a–machine* contains 380 *states*. Next, the supervisory structure is obtained as the conditionally controllable *a–machine* where all uncontrollable transitions are allowed to generate conditionally controllable behaviour. The obtained supervisory structure contains 110 *states* with 19 of them identified as potential generators of uncontrollable behaviour. The corresponding *a–machine* is listed in tables 6.2, 6.3 and 6.4. The *states* and associated *state–variables* together with these transitions causing conditional controllability are listed in table 6.5.

S	11	12	13	14	15	16	21	22	23	24	25	26	27	28	31	32	33	34	35	36	37	38	41	42	51	52
1	4						2	3	79		81				87										71	
2									5																72	
3																									73	
4																										
5		6								98		53											57		38	
6			7							35		48											58		39	
7										8		49											59		40	
8																							9		41	
9				10	11									9										8	22	
10														10										35	23	
11														11										98	24	
12		13	14											98										65	25	
13				13										35										64	26	
14		16												8										63	27	
15			15		17								17	22										20	28	
16				16									16										17	19	29	
17		19											15										16	18	30	
18																							15		31	
19			20																				29		32	20
20																							23		33	
21																									21	
22																										8
23				24										23										22		9
24					25									24										34		10
25														25										36		11
26		27			26								31	36										60		12
27			28	27									30	34										61		13
28													29											62		14
29			29	30																				21		15
30					31																		31	33		16
31		30																					30	32		17
32		33																					24			18
33			21																				10			19
34																							25			
35																									34	35
36							37								67											98

Table 6.2: *a-machine* corresponding to the supervisory structure (part A)

S	11	12	13	14	15	16	21	22	23	24	25	26	27	28	31	32	33	34	35	36	37	38	41	42	51	52
								TRANSITIONS																		
37								38			60															66
38									39																	4
39		40								36		46											56			5
40			41							34		47											55			6
41										22		42											54			7
42													21										43			49
43				44									29											42		50
44		44	43		45								30										45	47		51
45		47											31										44	46	47	52
46			42										32										51		42	53
47			49										33										50		43	48
48													19												44	
49				51									20												45	
50		51	50		52								15											49	46	
51		48											16											48		
52													17											53		
53													18										52			
54				55						23		43												41	56	59
55		55	54		56					24		44												40	55	58
56		58		58						25		45												39	54	57
57			59		57					11		52												5		
58										10		51												6		
59										9		50												7		
60		61											32										26			65
61			62										33										27		62	64
62													21										28		61	63
63													20										14		60	
64		64	63										19										13		37	
65								4			65		18										12			
66																										
67						69										68										99
68																	92									100
69																	70									88
70																			110							89
71								72			80				82			71		91						1
72								73																		2

Table 6.3: *a–machine* corresponding to the supervisory structure (part B)

S	11	12	13	14	15	16	21	22	23	24	25	26	27	28	31	32	33	34	35	36	37	38	41	42	51	52
73	38								74																	3
74	39									71				71												79
75	46											75		1											76	78
76	32												76												75	77
77	18																								74	
78	53												77													
79	5									1		78														81
80	60												76												80	
81	65												77													
82																69			83		84	71			84	87
83																						1			83	86
84																	89		86		85				82	85
85																88									69	
86																									70	
87																									91	
88							66																			
89																		1		90						
90																										
91						70															85		106			90
92						91												36		93	84		104		95	101
93						84															94		95		94	102
94																						36		94	36	97
95						85																25		97	67	96
96																						11			68	
97						88																98	96		92	
98						89									99				107				11		93	
99						90										100									104	
100																	101	98		102			105			
101																					97		103	102		
102																					96			93		
103																		11		103	95			101		
104																		25		104				92		
105																										
106						86																	108		106	103
107																					97			107		
108																					96			110	110	105
109																					95				109	108
110						83															94		109			107

TRANSITIONS

Table 6.4: *a–machine* corresponding to the supervisory structure (part C)

S	M	State variables					Cnd. ctrbl.	Unct. trans.
1	*	0	0	0	0	0		
2		0	1	0	0	0		
3		0	2	0	0	0		
4		1	2	0	0	0	*	12
5		1	3	0	0	0		
6		2	3	0	0	0		
7		3	3	0	0	0		
8		3	0	0	0	0		
9		3	0	0	1	0		
10		2	0	0	1	0		
11		1	0	0	1	0	*	16
12		1	5	0	1	0	*	16
13		2	5	0	1	0		
14		3	5	0	1	0		
15		3	6	0	1	0		
16		2	6	0	1	0		
17		1	6	0	1	0	*	16
18		1	6	0	0	0		
19		2	6	0	0	0		
20		3	6	0	0	0		
21		3	6	0	0	1		
22		3	0	0	0	1		
23		3	0	0	1	1		
24		2	0	0	1	1		
25		1	0	0	1	1	*	16
26		1	5	0	1	1	*	16
27		2	5	0	1	1		
28		3	5	0	1	1		
29		3	6	0	1	1		
30		2	6	0	1	1		
31		1	6	0	1	1	*	16
32		1	6	0	0	1		
33		2	6	0	0	1		
34		2	0	0	0	1		
35		2	0	0	0	0		
36	*	1	0	0	0	1		
37		1	1	0	0	1	*	16
38		1	2	0	0	1		
39		1	3	0	0	1		
40		2	3	0	0	1		
41		3	3	0	0	1		
42		3	4	0	0	1		
43		3	4	0	1	1		
44		2	4	0	1	1		
45		1	4	0	1	1	*	16
46		1	4	0	0	1		
47		2	4	0	0	1		
48		2	4	0	0	0		
49		3	4	0	0	0		
50		3	4	0	1	0		
51		2	4	0	1	0		
52		1	4	0	1	0	*	16
53		1	4	0	0	0		
54		3	3	0	1	1		
55		2	3	0	1	1		
56		1	3	0	1	1	*	16
57		1	3	0	1	0	*	16
58		2	3	0	1	0		
59		3	3	0	1	0		
60		1	5	0	0	1		
61		2	5	0	0	1		
62		3	5	0	0	1		
63		3	5	0	0	0		
64		2	5	0	0	0		
65		1	5	0	0	0		
66		1	1	0	0	0		
67		1	0	1	0	1		
68		1	0	2	0	1		
69		0	0	2	0	1		
70		0	0	3	0	1		
71	*	0	0	0	0	1		
72		0	1	0	0	1		
73		0	2	0	0	1		
74		0	3	0	0	1		
75		0	4	0	0	1		
76		0	6	0	0	1		
77		0	6	0	0	0		
78		0	4	0	0	0		
79		0	3	0	0	0		
80		0	5	0	0	1		
81		0	5	0	0	0		
82		0	0	1	0	1		
83		0	0	5	0	1		
84		0	0	6	0	1		
85		0	0	6	0	0		
86		0	0	5	0	0		
87		0	0	1	0	0		
88		0	0	2	0	0		
89		0	0	3	0	0		
90		0	0	4	0	0		
91		0	0	4	0	1		
92		1	0	3	0	1		
93		1	0	4	0	1		
94		1	0	6	0	1		
95		1	0	6	1	1		
96		1	0	6	1	0	*	16
97		1	0	6	0	0		
98	*	1	0	0	0	0		
99		1	0	1	0	0		
100		1	0	2	0	0		
101		1	0	3	0	0		
102		1	0	4	0	0		
103		1	0	4	1	0	*	16
104		1	0	4	1	1	*	16
105		1	0	3	1	0	*	16
106		1	0	3	1	1	*	16
107		1	0	5	0	0		
108		1	0	5	1	0	*	16
109		1	0	5	1	1	*	16
110		1	0	5	0	1		

Table 6.5: *State-variable* values for states of the *a-machine* corresponding to the supervisory structure of the metering system.

Controller	Number of states	Uncontrollable transitions causing conditional controllable behaviour
N1	110	none
N2	110	none
N3a	112	51, 52
N3b	108	51, 52
N4	112	51, 52
N5a	110	none
N5b	165	51
A6a	118	13, 24, 26, 28, 51, 52
A6b	120	12, 24, 26, 28, 51, 52
A7a	110	11, 12, 13, 14, 15, 51, 52
A7b	100	11, 12, 51, 52
A8a	110	16, 51, 52
A8b	108	16, 51, 52

Table 6.6: Local controllers. Size and uncontrollable transitions causing conditionally controllable behaviour.

6.3.5 Step 4.- Controller Synthesis

Once the TL formulas representing the dynamic specifications are translated into *a–machines*, a local controller is synthesised for each using the supervisory structure as a base model. All uncontrollable transitions are considered as potential causes of conditionally controllable behaviour. Thus, the only trajectories eliminated in each local controller are those not existing in the supervisory structure. The size of each local controller is shown in table 6.6, together with the list of uncontrollable transitions causing conditionally controllable behaviour with respect to the supervisory structure in each local controller.

The global controller is calculated obtaining the *controller conjunction* of all local controllers. The resultant structure has 101 *states*, from which 67 conditionally controllable behaviour with respect to the supervisory structure can occur. It is shown in tables 6.7, 6.8 and 6.9. Table 6.10 shows the *state–variables* values associated to each *state* of the structure indicating what *states* generate conditionally controllable behaviour and what are the uncontrollable transitions causing it.

TRANSITIONS

S	11	12	13	14	15	16	21	22	23	24	25	26	27	28	31	32	33	34	35	36	37	38	41	42	51	52
1																										100
2		5																								97
3							3																		2	
4								4	91		99															
5									6																	
6		7								13		88											89			75
7																							8			
8			9		84					11		85												86		73
9				8						10		51												68		72
10				11	12																			48		50
11																								44		46
12																								13		
13														14												
14						16									15											31
15																24			43							32
16																	17	2								
17																				18						22
18																					19					
19																						2				20
20																						21				
21																										
22																		21		23	20					
23																										
24						17														25			41			34
25																		13			26					
26						19																				
27																						13	27	26		36
28																						28		13		37
29																								30		29
30																			40							
31						33									31											
32															32	34										
33																	22									

Table 6.7: *a–machine* corresponding to the global controller of the metering system (part A).

S	11	12	13	14	15	16	21	22	23	24	25	26	27	28	31	32	33	34	35	36	37	38	41	42	51	52
34						22												30		35			38			
35																										
36						20															36			36		
37																								34		38
38																					37	30				
39																		29		39	36	29				
40																										
41																		28		42	27			24		45
42																					26					
43																										
44																							11			
45																							46	45		49
46					47																			30		
47																										
48																							10			
49																							50	49		
50				46																						
51														10										67		60
52			52										52	11										66		59
53				53	54																			55		
54														13												63
55		56																						55		
56																								63		58
57		53												28									57	65		
58		59												29									53	61		
59														46												
60			60	59	62									50									60	63		
61														49									58			
62		64																								
63														30												
64																							59			
65			61											45									59			
66			67											44									53			65
67														48									52			61

Table 6.8: *a-machine* corresponding to the global controller of the metering system (part B).

TRANSITIONS

S	11	12	13	14	15	16	21	22	23	24	25	26	27	28	31	32	33	34	35	36	37	38	41	42	51	52
68										48		69											9			70
69										49		71	67													
70																							72			
71													61													
72			72							50		83												70		
73				73						46		80												81		
74					74																			75		
75		76								30		77											78			
76																							73			
77													63													
78		73								29		79												75		
79													58													
80													59													
81			70							45		82											73			
82													65													
83													60													
84																								6		
85													53													
86			68							44		87											8			81
87													66													
88													55													
89		8								28		90												6		78
90													57													
91	6									2		92	93													95
92												93			2											
93	55														21											94
94	63																									
95	75									21		96	94													
96									95																	
97	98								75																	
98								97				93														
99											101															
100												94														
101																										

Table 6.9: *a–machine* corresponding to the global controller of the metering system (part C).

S	M	State variables					Cnd. ctrbl.	Unct. trans.
1	*	0	0	0	0	0		
2	*	0	0	0	0	1	*	62
3		0	1	0	0	1		
4		0	2	0	0	1		
5		1	2	0	0	1	*	62
6		1	3	0	0	1		
7		2	3	0	0	1	*	13, 24, 26, 62
8		2	3	0	1	1		
9		3	3	0	1	1		
10		3	0	0	1	1		
11		2	0	0	1	1		
12		1	0	0	1	1	*	62
13	*	1	0	0	0	1	*	62
14		1	0	1	0	1		
15		1	0	2	0	1		
16		0	0	2	0	1	*	62
17		0	0	3	0	1		
18		0	0	4	0	1	*	62
19		0	0	6	0	1		
20		0	0	6	0	0	*	61
21	*	0	0	0	0	0	*	61
22		0	0	3	0	0	*	61
23		0	0	4	0	0	*	61
24		1	0	3	0	1		
25		1	0	4	0	1	*	16, 62
26		1	0	6	0	1		
27		1	0	6	1	1		
28		1	0	0	1	1		
29		1	0	0	1	0	*	61
30	*	1	0	0	0	0	*	61
31		1	0	1	0	0	*	61
32		1	0	2	0	0	*	61
33		0	0	2	0	0	*	61
34		1	0	3	0	0	*	61
35		1	0	4	0	0	*	16, 61
36		1	0	6	0	0	*	61
37		1	0	6	1	0	*	61
38		1	0	3	1	0	*	61
39		1	0	4	1	0	*	61
40		1	0	5	0	0	*	16, 61
41		1	0	3	1	1		
42		1	0	4	1	1	*	62
43		1	0	5	0	1	*	16, 62
44		2	0	0	0	1		
45		2	0	0	0	0	*	61
46		2	0	0	1	0	*	61
47		1	0	0	1	0	*	61
48		3	0	0	0	1		
49		3	0	0	0	0	*	61
50		3	0	0	1	0	*	61
51		3	4	0	1	1	*	14, 62
52		3	6	0	1	1		
53		2	6	0	1	1		
54		1	6	0	1	1	*	12, 28, 62
55		1	6	0	0	1		
56		2	6	0	0	1	*	13, 28, 62
57		1	6	0	1	1		
58		1	6	0	1	0	*	12, 28, 61
59		2	6	0	1	0	*	61
60		3	6	0	1	0	*	61
61		3	6	0	0	0	*	61
62		1	6	0	1	0		
63		1	6	0	0	0	*	61
64		2	6	0	0	0	*	13, 28, 61
65		2	6	0	0	0		
66		2	6	0	0	1		
67		3	6	0	0	1		
68		3	3	0	0	1		
69		3	4	0	0	1	*	62
70		3	3	0	0	0	*	61
71		3	4	0	0	0	*	61
72		3	3	0	1	0	*	61
73		2	3	0	1	0	*	61
74		1	3	0	1	0	*	12, 24, 26, 61
75		1	3	0	0	0	*	61
76		2	3	0	0	0	*	13, 24, 26, 61
77		1	4	0	0	0	*	12, 61
78		1	3	0	1	0		
79		1	4	0	1	0	*	12, 61
80		2	4	0	1	0	*	13, 15, 61
81		2	3	0	0	0		
82		2	4	0	0	0	*	13, 61
83		3	4	0	1	0	*	14, 61
84		1	3	0	1	1	*	12, 24, 26, 62
85		2	4	0	1	1	*	13, 15, 62
86		2	3	0	0	1		
87		2	4	0	0	1	*	13, 62
88		1	4	0	0	1	*	12, 62
89		1	3	0	1	1		
90		1	4	0	1	1	*	12, 62
91		0	3	0	0	1		
92		0	4	0	0	1	*	11, 62
93		0	6	0	0	1		
94		0	6	0	0	0	*	61
95		0	3	0	0	0	*	61
96		0	4	0	0	0	*	11, 61
97		0	2	0	0	0	*	61
98		1	2	0	0	0	*	61
99		0	5	0	0	1	*	11, 62
100		0	1	0	0	0	*	61
101		0	5	0	0	0	*	11, 61

Table 6.10: *State-variable* values for *states* of the *a-machine* corresponding to the global controller of the metering system.

6.3.6 Controller Implementation and Simulation Results

A discrete/continuous model of the metering tank is implemented using the
same g*PROMS* model of the buffer system, with the addition of an extra level
switch and an outlet connection for valve RV. This valve is modelled by the
s_valve model of fig. 6.13. FV and EV are modelled using an extended *s_valve*
model which considers the valve getting stuck open or closed or being under
repair as shown in figs. 6.29 and 6.30. The failure rates in opening and closing
the valves follow a normal distribution. If the value of the normal distribution
variable is between certain preset minimum and maximum values, the valve
executes the operation successfully. Outside those boundaries, the valve fails.
The on/off switch is modelled with a **SELECTOR** variable to describe its
status. Its model is shown in fig 6.31. Tank, valves and switch models are
instantiated to form the *tank* model shown in fig 6.32.

The controller synthesised in th previous section was automatically translated into g*PROMS* code following the same procedure as in the previous example. Typical control tasks are shown in figs. 6.33 and 6.34. Task Check_state_13
in fig. 6.33, issues the command to open valve EV. Task Check_state_14 in
fig. 6.34 monitors the system response. This may consist of the indication
that EV is open, EV is stuck closed or the on/off switch is changing its position to off. Each of these responses make the controller switch to controller
states 15, 43 and 31 respectively. If no response occurs after MAX_SAMPLE_TIME
units have elapsed, the controller reaches its failure state (*state* 102). Task
Check_state_15 in fig. 6.34 also monitors the system response. If level is
detected empty, the controller switches to *state* 16 in which the command to
close EV will be issued (see table 6.7). Another possible system response is
that the on/off switch is turned off, which will switch the system to *state* 32.
Note that the command to close EV can be issued in this control task if the
sample time is bigger than MAX_WAIT_TIME. The existence of this controllable
transition in the current control task is because in the specification modelling
section no specification was given prohibiting the execution of such a transition
under these circumstances.

A typical result of the controller driving the system is shown in fig. 6.35.
Relevant parameters of this simulation are shown in table 6.11. The proba-
bilities to open and close successfully FV and EV valves were defined so low
for the purposes of testing the controller. The upper graph shows the level
measurement profile. FV, EV and RV flowrates are shown in the middle graph
and a profile of the controller state number is shown in the lower graph.

The on/off switch is turned on (controller *state* 2) at $t = 5$ and operation starts. FV succeeds in opening at the first attempt and the tank starts building up volume. At $t = 16.1$ the level measurement signals to be within discharge range (controller *state* 4). The controller sends the command to close FV (controller *state* 5) at $t = 17$ but at $t = 17.5$ the process emits the signal that the level has gone beyond the discharge range. FV closes at $t = 18$ (controller *state* 6) and, because the level must be readjusted, at $t = 19$ RV opens (controller *state* 7). At t= 20.8 the level measurement reaches the discharge range again (controller *state* 8) and RV is closed at $t = 22$ (controller *state* 84). At $t = 23$, the controller issues the command to open EV (controller *state* 13). It is detected open at $t = 24$ (controller *state* 14) and at $t = 25$ the controller is ready to receive the signal from the level measurement device. Liquid is discharged until the response of minimum level is produced at $t = 54.3$ The controller accepts the signal at $t = 55$ (controller *state* 16). The controller then sends the command to close EV (controller *state* 17) and at $t = 57$ it receives back the system response that EV is stuck open (controller *state* 18). The controller sends the repair command (controller *state* 19) and at $t = 62$ it receives the response that the valve has been repaired and is closed. The controller returns to *state* 2.

At this stage, the first operation cycle is completed. The controller sends the command to open FV to start again the cycle at $t = 63$ (controller *state* 3). At $t = 64$ FV fails to open (controller *state* 99). The controller sends the repair command (controller *state* 93) at $t = 65$. FV finishes its repair cycle at $t = 75$. The controller sends the command to open FV at $t = 66$ but it accepts the response of the system at $t = 77$ that the valve has failed again. The controller issues again the repair command at $t = 78$. FV finishes its repair cycle at $t = 88$. The controller attempts to restart operation and succeeds this time. At $t = 98.5$, the level device issues the response indicating that is in discharge range and the same situation as in the previous cycle is repeated. That is, the level is building up so fast that before FV is detected closed the level measurement has gone beyond the discharge level and has to be corrected using RV. At $t = 106$ the level has been already adjusted and RV has been closed. The controller sends the command to open EV. It fails to open (controller *state* 43) and the controller sends the command to repair it and waits until the repair operation is finished. After that, the controller tries again to open EV and succeeds. At $t = 142.7$ the minimum level signal is issued by the process and the controller accepts it at $t = 143$. The controller orders the closing of EV at $t = 144$. At $t = 145$, the controller detects EV closed, finishing the second cycle. The controller starts the third cycle sending the command

PARAMETER	VALUE
MAX_SAMPLE_TIME	207 time units
MAX_WAIT_TIME	203 time units
SIMULATION_TIME	315 time units
surge_tank.DENSITY	1000 mass/volume units
surge_tank.AREA	1 area units
surge_tank.MAXVOL	30 volume units
surge_tank.NORMALVOL	15 volume units
surge_tank.HIGHVOL	20 volume units
surge_tank.EMPTYVOL	1 volume units
surge_tank.VOL_TOL	0.5 volume units
FV valve.MAX_REPAIR_TIME	10 time units
EV valve.MAX_REPAIR_TIME	4 time units
on/off switch.START_TIME	5 time units
on/off switch.END_TIME	270 time units
P(valve open successfully)	0.43
P(valve close successfully)	0.34

Table 6.11: Relevant parameters for simulation shown in fig 7.33.

to open FV. In this cycle, FV fails once to open while EV fails once to open and once to close. The fourth cycle starts at $t = 220$. This time, FV succeeds in opening without any failure and at $t = 230.5$ discharge range level is issued. The controller orders to close FV but again the level is increasing too quickly and RV is opened at $t = 234$ to readjust the level. Meanwhile, at $t = 235$, FV is detected stuck open for the first time during the simulation, causing the level to increase while RV drains liquid continuously to the dump tank. The repair command is sent at $t = 237$. The repair of FV ends at $t = 247$. RV stays open until the level measurement is within discharge range and closed at $t = 255$. Then the controller orders EV to open but it fails twice. While trying to open EV for the second time, the on/off switch is turned off at $t = 270$. The controller detects EV open at $t = 271$ and at the next time instant the on/off switch is detected in the off position. Minimum level signal is issued at $t = 299.9$ and EV closes at the first attempt reaching the final state (state 21) at t = 302, terminating the operation. Overall, the controller visits 28 different states out of 102 within the gPROMS implementation.

```
#
#========================================================
#
# Model of a solenoid valve plus failure states
#
#========================================================
#
 MODEL s`comp`valve
 PARAMETER
   VALVE`CONSTANT        AS REAL
   MAX`REPAIR`TIME       AS REAL
 VARIABLE
   position              AS open`fraction
   flow                  AS flow`rate
   press`in              AS pressure
   press`out             AS pressure
   delta`p               AS pressure
   c`action              AS NoType
   r`action              AS NoType
   failure`rate          AS NoType
   repair`time`init      AS NoType
 SELECTOR
   s`comp`valve`status   AS (closed, closing, open, opening,
                            stuck`clsd, stuck`opn,
                            repair`open, repair`clsd)
                            DEFAULT closed
 STREAM
   input  : flow, press`in   AS Mainstream
   output : flow, press`out AS Mainstream
 SET
   VALVE`CONSTANT := 40 ;
 EQUATION
 # Flowrate / Delta P relationship
   flow = position*VALVE`CONSTANT*SGN(delta`p)*SQRT(ABS(delta`p)) ;
 # Delta`P definition
   delta`p = press`in - press`out ;
   CASE s`comp`valve`status OF
     WHEN closed    : position = 0.0;
                      SWITCH TO opening    IF c`action    = 21 ;
     WHEN opening   : position = 0.0;
                      SWITCH TO open          IF failure`rate GT 0.3 and
                                               failure`rate LT 0.7 ;
                      SWITCH TO stuck`clsd IF failure`rate LE 0.3 or
                                               failure`rate GE 0.7 ;
     WHEN open      : position = 1.0;
                      SWITCH TO closing    IF c`action    = 23 ;
     WHEN closing   : position = 1.0;
                      SWITCH TO closed        IF failure`rate GT 0.3 and
                                               failure`rate LT 0.7 ;
                      SWITCH TO stuck`opn   IF failure`rate LE 0.3 or
                                               failure`rate GE 0.7 ;
     WHEN stuck`opn : position = 1.0;
                      SWITCH TO repair`opn IF c`action    = 27 ;
     WHEN stuck`clsd : position = 0.0;
                      SWITCH TO repair`clsd IF c`action   = 27 ;
     WHEN repair`clsd: position = 0.0;
                      SWITCH TO closed        IF (TIME-repair`time`init) GE
                                               MAX`REPAIR`TIME ;
     WHEN repair`opn : position = 1.0;
                      SWITCH TO closed        IF (TIME-repair`time`init) GE
                                               MAX`REPAIR`TIME ;
   END # case
 END # model s`comp`valve
```

Figure 6.29: *gPROMS* generic model of an on/off valve with failure states.

```
#
#===========================================================
#
# Task to open a s`comp`valve
#
#===========================================================
#
 TASK open`s`comp`valve
PARAMETER
  s`comp`valve            AS MODEL s`comp`valve
SCHEDULE
   PARALLEL
     RESET s`comp`valve.failure`rate := NORMAL(0.5, 0.35) ; END
     RESET s`comp`valve.c`action     := 21 ; END
   END # parallel
END # task open`s`comp`valve
#
#===========================================================
#
# Task to close a s`comp`valve
#
#===========================================================
#
 TASK close`s`comp`valve
PARAMETER
  s`comp`valve            AS MODEL s`comp`valve
SCHEDULE
   PARALLEL
     RESET s`comp`valve.failure`rate := NORMAL(0.5, 0.44) ; END
     RESET s`comp`valve.c`action     := 23 ; END
   END # parallel
END # task close`s`comp`valve
#
#===========================================================
#
# Task to repair a s`comp`valve
#
#===========================================================
#
 TASK repair`s`comp`valve
PARAMETER
  s`comp`valve            AS MODEL s`comp`valve
SCHEDULE
   PARALLEL
     RESET s`comp`valve.repair`time`init := OLD(TIME) ; END
     RESET s`comp`valve.c`action     := 27 ; END
   END # parallel
END # task close`s`comp`valve
```

Figure 6.30: *gPROMS* generic tasks for the operation of an on/off valve with failure states.

```
#
#==========================================================
#
# Model of an on-off button
#
#==========================================================
#
MODEL button
PARAMETER
  START'TIME                        AS REAL
  END'TIME                          AS REAL
VARIABLE
  position                          AS Open'fraction
SELECTOR
  button'status                     AS (stat'on, stat'off) DEFAULT stat'off
EQUATION
  CASE button'status OF
    WHEN stat'on        : position = 1.0;
                        SWITCH TO stat'off       IF TIME GE END'TIME ;
    WHEN stat'off       : position = 0.0;
                        SWITCH TO stat'on        IF (TIME GE START'TIME) and
                                                 (TIME LT END'TIME);
    END # case
END # model button
```

Figure 6.31: *gPROMS* model of the on/off switch.

```
#
#==========================================================
#
# Model the tank plus valves plus button
#
# =========================================================
#
MODEL tank
VARIABLE
  sample'time                       AS time'counter
PARAMETER
  ATM'PRESS                         AS REAL
  MAX'WAIT'TIME                     AS REAL
  MAX'SAMPLE'TIME                   AS REAL
  SIMULATION'TIME                   AS REAL
UNIT
  fill'valve                        AS s'comp'valve
  exit'valve                        AS s'comp'valve
  drain'valve                       AS s'valve
  surge'tank                        AS atm'tank
  button                            AS button
EQUATION
# STREAM
  fill'valve.output        IS   surge'tank.input  ;
  exit'valve.input         IS   surge'tank.output ;
  drain'valve.input        IS   surge'tank.drain  ;
END # model tank
```

Figure 6.32: *gPROMS* model of complete metering system.

```
# ==========================================================
#
# Task state 13
#
# ==========================================================
#
TASK Check'state'13
PARAMETER
  fsa          AS MODEL fsa
  tank         AS MODEL tank
SCHEDULE
  IF (fsa.state = fsa.s13) and (fsa.detect = 0) THEN
    SEQUENCE
      open's'comp'valve(s'comp'valve IS tank.exit'valve);
      RESET fsa.detect := 1 ; END
      SWITCH fsa.state := fsa.s14; END
    END # sequence
  END # if
END # task
#
# ==========================================================
#
# Task state 14
#
# ==========================================================
#
TASK Check'state'14
PARAMETER
  fsa          AS MODEL fsa
  tank         AS MODEL tank
SCHEDULE
  IF (fsa.state = fsa.s14) and (fsa.detect = 0) THEN
    SEQUENCE
      IF (tank.exit'valve.s'comp'valve'status = tank.exit'valve.open)
        and (fsa.detect = 0) THEN
        SEQUENCE
          RESET tank.sample'time := OLD(TIME) ; END
          RESET fsa.detect := 1 ; END
          SWITCH fsa.state := fsa.s15; END
        END # sequence
      END # if
      IF (tank.exit'valve.s'comp'valve'status = tank.exit'valve.stuck'closed)
        and (fsa.detect = 0) THEN
        SEQUENCE
          RESET tank.sample'time := OLD(TIME) ; END
          RESET fsa.detect := 1 ; END
          SWITCH fsa.state := fsa.s43; END
        END # sequence
      END # if
      IF (tank.button.button'status = tank.button.stat'off ) and (fsa.detect = 0) THEN
        SEQUENCE
          RESET tank.sample'time := OLD(TIME) ; END
          RESET fsa.detect := 1 ; END
          SWITCH fsa.state := fsa.s31; END
        END # sequence
      END # if
      IF (TIME - tank.sample'time GT tank.MAX'SAMPLE'TIME) and (fsa.detect = 0)THEN
        SEQUENCE
          RESET tank.sample'time := OLD(TIME) ; END
          RESET fsa.detect := 1 ; END
          SWITCH fsa.state := fsa.s102 ; END
        END # parallel
      END # if
    END # sequence
  END # if
END # task
```

Figure 6.33: *gPROMS* typical control tasks for the operation of the metering system (part A).

```
#
# ================================================================
#
# Task state 15
#
# ================================================================
#
TASK Check'state'15
PARAMETER
  fsa          AS MODEL fsa
  tank         AS MODEL tank
SCHEDULE
  IF (fsa.state = fsa.s15) and (fsa.detect = 0) THEN
    SEQUENCE
      IF (tank.surge'tank.level'switch =  tank.surge'tank.empty ) and (fsa.detect = 0) THEN
        SEQUENCE
          RESET tank.sample'time := OLD(TIME) ; END
          RESET fsa.detect := 1 ; END
          SWITCH fsa.state := fsa.s16; END
        END # sequence
      END # if
      IF (TIME - tank.sample'time GT tank.MAX'WAIT'TIME) and (fsa.detect = 0) THEN
        SEQUENCE
          close's'comp'valve(s'comp'valve IS tank.exit'valve);
          RESET fsa.detect := 1 ; END
          SWITCH fsa.state := fsa.s24; END
        END # sequence
      END # if
    END # sequence
  END # if
END # task
```

Figure 6.34: *gPROMS* typical control tasks for the operation of the metering system (part B).

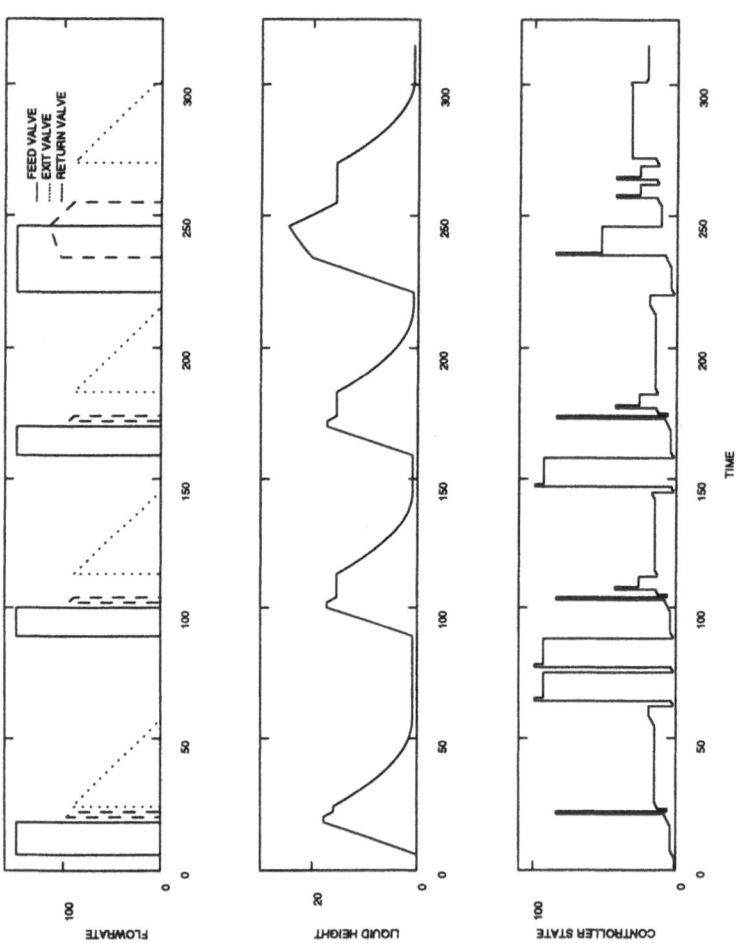

Figure 6.35: Typical simulation results for the metering system. The upper, middle an lower graphs show the flow rate of feed (FV), exit (EV) and return (RV) valves, the level profile and the controller state trajectories, respectively.

Chapter 7

Epilogue

A method has been proposed for the formal specification and synthesis of procedural controllers whose structure can be represented by labelled *FSMs* called *a−machines*. These structures are amenable to formal proofs of correctness. The method is based on

- A consistent modelling framework for process, behaviour specifications and structures for supervision and control.

- An extension of the Supervisory Control Theory, in which a broader class of process behaviour is considered to include systems of practical interest in process engineering. Control mechanisms are proposed which are capable of issuing controllable transitions to the process.

7.1 The Modelling Framework

The proposed method requires as input information a model of the process to be controlled and models of the behaviour to be imposed upon the process. Due to their inherent characteristics, they are modelled using different frameworks. The process is modelled using *FSMs* while specification are modelled predicate and temporal logic formalisms.

7.1.1 Process Modelling

Labelled finite state machines, called *a−machines*, are the basic modelling tool in which each *state* is characterised by a set of *state−variables* describing the process. These *state−variables* take values from finite domains. Two extra symbols, the covering (∞_j) and the partial covering ($\infty_j^{Le_j}$), are included to handle incomplete information in the model and to facilitate its construction. The covering symbol can be assigned to any *state−variable j* about which no

information is available or if it is of no interest to the current *state*. The
partial covering symbol excludes from the possible values of *state–variable j*
those defined in the subset Le_j. The formal handling of these symbols is based
on the partial orders *covering* and *partial covering* on the *states* and *state-
variables*. They also introduce an ordering in the *state–variables* and *states*
resulting in compact models. Moreover, they establish part of the formal link
for the mapping of behaviour specifications modelled by logic formalisms into
the *a–machine* domain in which the process model construction and controller
synthesis is performed. The model construction was illustrated with several
examples in which the advantages of the use of the covering symbols and partial
orders were demonstrated.

The modelling framework permits the handling of states and/or transitions
as required using a well structured and compact representation. This is a necessary characteristic when dealing with realistic models involving calculations
with extensive manipulation of states. Recently, Li and Wonham (1993) also
recognised that in order to deal with more realistic cases, a more powerful modelling tool than that proposed in the Supervisory Control Theory was needed.
They extended the basic *FSM*–based model of the Supervisory Control Theory to a vector representation based on vector additions such as those used
in the evolution of markings of Petri Nets. The "state" of a vector DES is
described by "state–variables" in the same fashion as in the modelling framework proposed here, but taking values from the integer domain $\mathcal{Z}$. The partial
function δ is extended to consider changes in the vector of state–variables representing each state. Each transition can consider the change of more than one
"state–variable".

Although state explosion is still an issue, efficient implementations have
been published which reduce the impact of computational complexity in the
handling of state–transition structures. Balemi *et al.* (1993) presented an efficient implementation based on Binary Decision Diagrams (Brace *et al.*, 1990).
This calculation technique has been also used by Burch *et al.* (1990) for the
model checking of structures of up to 10^{20} states. Moreover, the use of well
structured models with defined algebraic operations such as the *a–machines*
allows the handling of process behaviour explicitly with only a relatively small
number of states. This is particularly useful in the construction of specification
models, as was shown in chapter 3.

7.1.2   Specification Modelling

Behaviour specification modelling has proved to be a difficult and error–prone activity, in particular for dynamic behaviour specifications. In order to overcome some of the problems, specifications stated in natural language are modelled using logic formalisms, thus eliminating ambiguities in their interpretation. Logic invariant specifications, in this case restricted only to forbidden states, are modelled as Predicate Logic statements. Specifications involving dynamic behaviour are modelled using TL formulas.

Forbidden states are translated into *a–machine states* using the method proposed by Ramadge and Wonham (1987a). This treatment could be extended to other logic–invariant properties such as buffer capacities or finite counting, using the same approach (Ramadge and Wonham, 1987a; Kumar *et al.*, 1991b).

In the case of TL formulas, these are constructed using a restricted syntax defined by three TL schemes for which homomorphisms to map formulas into the *a–machine* domain have been developed. These homomorphisms give sufficient conditions to guarantee the existence of an *a–machine* for any TL formula which is valid in the TL frame given by the *asynchronous product* of the elementary process models. The algebraic properties associated with the *a–machines* are exploited in the translation process resulting in models in the *a–machine* domain that can be used directly in the controller synthesis.

TL was favoured over other Temporal Logic frameworks because of its relative simplicity and previous application in the analysis of dynamic systems. Thistle and Wonham (1986) made a first adaptation of the proof system of Manna and Pnueli (1983). This was later improved and incorporated in the RTTL/ESM system of Ostroff (1989b) for analysis of real–time systems. Although no real applications of the RTTL/ESM system have been reported, the proof system and the decision methods proposed to determine if a TL formula is true could be exploited in the construction of dynamic specifications and their translation into the *a–machine* domain. Currently, the type of TL formulas used to model behaviour specification is restricted to the syntax of the three schemes defined. The number of schemes could be extended as necessary, by developing corresponding homomorphisms so as to characterise a wider behaviour of process systems. TL formulas modelling more complex behaviour could then be constructed using the complete TL syntax and then reduced to a combination of the basic schemes using the proof system from which the translation into the *a–machine* domain could be executed. Well developed tools from software engineering could be used to construct and automate the proof procedures (Jones *et al.*, 1991).

An important issue not addressed here is the analysis of behaviour specifications. The identification of redundant or contradictory specifications, or the determination of whether a group of behaviour specifications is a "complete" set for a given process are important questions to be answered for which logic tools could be useful.

7.2   Supervision and Control Aspects

The developed framework for the synthesis of procedural controllers is based on the controllability and supervision aspects of the Supervisory Control Theory proposed by Ramadge and Wonham (1987a; 1987b) and Wonham and Ramadge (1987). However, it was shown here that the concept of controllability originally proposed makes the synthesis of supervisors (and controllers) difficult and even rules out behaviour of interest for process systems. To overcome this limitation, the following additions were made:

- Proposal of conditional controllability to allow the existence of uncontrollable transitions from a given state that need not be executed.

- Definition of supervisors that meet conditional controllability and development of sufficient conditions for their existence.

- Development of a calculation method for conditional controllable behaviour extending the method proposed by Wonham and Ramadge (1987) for closed languages.

Moreover, Supervisory Control Theory was conceived only to provide mechanisms which maintain the closed–loop process behaviour within a finite set of trajectories by disabling controllable actions. In that respect, the closed–loop behaviour can be considered as a superset of controllable trajectories. However, no mechanisms were proposed for the enforcement of given controllable transitions to satisfy specific control objectives. In order to achieve this, two novel concepts of controller were introduced based on a different interpretation of the structure realising the closed–loop behaviour for a given system:

- *Local controller*, a structure describing the enforcement of behaviour prescribed by a dynamic specification.

- *Global controller*, as the *controller conjunction* of all the *local controllers*, mapping each system state into at most one controllable transition.

The closed–loop behaviour realised by a local controller C is defined as the intersection of the marked and un-marked behaviours realised by the controller

C and the supervisory structure S satisfying static specifications of a given process, in which only strings from the supervisory structure that are specified by the controller survive. In this way it is guaranteed that the local controller will generate a closed–loop behaviour that not only satisfies the given dynamic specification but also avoids all the forbidden states.

Having obtained a *local controller* for each specification, a *global controller* is obtained as the *controller conjunction* of all local controllers. In order to guarantee the existence of a *global controller*, the marked behaviour of all trim *local controllers* in the *conjunction* must be nonconflicting. If the resultant structure contains *states* in which more than one controllable transition exist, extra dynamic specifications are included. These must eliminate all but one of the controllable transitions from each controller *state*. At present no method is proposed to carry out this specification refinement process automatically. Moreover, conditions of existence for a global controller were not defined. This is a topic that must be explored further. Besides, extra information concerning the nature of transitions may also be required because when dealing with the synthesis of controllers for more realistic cases, it was found that the classification of transitions as controllable or uncontrollable gives a poor characterisation of the process behaviour. For instance, some uncontrollable transitions are part of the normal behaviour of the process while others reflect abnormal conditions. For example, the detection of a high level which triggers the opening of a valve is part of the normal operation of the metering tank. The response of a valve indicating that is stuck open leads clearly to an emergency procedure to cope with abnormal operating conditions. Moreover, the execution of the same transition in different circumstances can lead to completely different behaviour. This was the case for the flame detection in the burner example. This reveals the need for a richer classification of transitions, which will help in the definition of conditions for the existence of global controllers and their calculation.

In principle, there may be several controllers which meet a set of given specifications. It thus become important to select one of the alternatives which is "best" according to some measure. Mathematical programming tools may be helpful for the systematic calculation of the global controller as has been shown by Li and Wonham (1994) and Yamalidou and Kantor (1991). The former employed integer programming methods for the calculation of an "optimal" supervisor while the latter addressed the synthesis of optimal marking for controlled Petri Nets.

7.3   The Controller Synthesis Method

Using the tools developed in chapter 2, 3 and 4, a controller synthesis method
was proposed based on an input–output interpretation of the process. The
method was divided into four steps:

- Step 1.- Process modelling.

- Step 2.- Specification modelling.

- Step 3.- Supervisory structure synthesis.

- Step 4.- Procedural Controller synthesis.

Four examples demonstrated its use and illustrated the features discussed
for the modelling framework, the modifications to the Supervisory Control Theory, the introduction of the notion of controller as well as important limitations
of the approach. The first two examples are based on the burner example presented by Moon *et al.* (1992), in which model checking techniques were used to
identify faults in the control logic for the startup and shutdown of the burner.
The examples emphasise

- The usefulness of the conditional controllability concept identifying sources
  of abnormal behaviour.

- The problems in the synthesis of a controller due to an "incomplete" set
  of behaviour specifications and the need to systematise the generation of
  a "closed" set of specifications. The set of dynamic specifications initially
  defined for the first example gave rise to a structure not satisfying the
  definition of global controller. Based on the obtained results, additional
  behaviour specifications were included. A key advantage of the proposed
  method is that it systematically highlights all possible sources of uncontrollable behaviour in the model. This facilitated the introduction of the
  required specifications for the global controller synthesis.

- The incremental construction of a controller to satisfy normal and abnormal operation conditions.

The latter two examples were based on a tank system with an associated
dynamic model describing "continuous" variables such as levels and flowrates.
These were implemented in the g*PROMS* dynamic simulator (Barton and Pantelides, 1994) together with the synthesised controllers to explore time–related
issues and to show the controllers performance. The examples highlighted the
following aspects:

- The rapid prototyping of behaviour specifications using logic formalisms.

- The feasibility of generating controllers for realistic problems.

- The structural uniformity of the synthesised controller. This will be of particular relevance when using formal verification methods.

- The importance of time related information to execute control actions. Pure *FSMs* do not consider time aspects. In the presence of more than one transition from a given controller *state*, the controller does not possess information to decide whether it must wait for the execution of an uncontrollable transition to occur or to execute the controllable transition, if present. In order to overcome this limitation, uncontrollable transitions were given priority of execution. This means that if a controllable transition must be executed from a given controller *state*, this must be specified as part of a dynamic specification in order to eliminate the remaining transitions from the appropriate controller *states*.

Appendix A

a–machines of Augmented Burner

A.1   Process Model

S	M	State variables	11	12	21	22	31	32	41	42	51	52	61	62
1	•	0,0,0,0,0,0	2		3		5		9		17		33	
2		1,0,0,0,0,0		1	4		6		10		18		34	
3		0,1,0,0,0,0	4			1	7		11		19		35	
4		1,1,0,0,0,0		3		2	8		12		20		36	
5		0,0,1,0,0,0	6		7			1	13		21		37	
6		1,0,1,0,0,0		5	8			2	14		22		38	
7		0,1,1,0,0,0	8			5		3	15		23		39	
8		1,1,1,0,0,0		7		6		4	16		24		40	
9		0,0,0,1,0,0	10		11		13			1	25		41	
10		1,0,0,1,0,0		9	12		14			2	26		42	
11		0,1,0,1,0,0	12			9	15			3	27		43	
12	•	1,1,0,1,0,0		11		10	16			4	28		44	
13		0,0,1,1,0,0	14		15			9		5	29		45	
14		1,0,1,1,0,0		13	16			10		6	30		46	
15		0,1,1,1,0,0	16			13		11		7	31		47	
16	•	1,1,1,1,0,0		15		14		12		8	32		48	
17	•	0,0,0,0,1,0	18		19		21		25			1	49	
18		1,0,0,0,1,0		17	20		22		26			2	50	
19		0,1,0,0,1,0	20			17	23		27			3	51	
20		1,1,0,0,1,0		19		18	24		28			4	52	
21		0,0,1,0,1,0	22		23			17	29			5	53	
22		1,0,1,0,1,0		21	24			18	30			6	54	
23		0,1,1,0,1,0	24			21		19	31			7	55	
24		1,1,1,0,1,0		23		22		20	32			8	56	
25		0,0,0,1,1,0	26		27		29			17		9	57	
26		1,0,0,1,1,0		25	28		30			18		10	58	
27		0,1,0,1,1,0	28			25	31			19		11	59	
28		1,1,0,1,1,0		27		26	32			20		12	60	
29		0,0,1,1,1,0	30		31			25		21		13	61	
30		1,0,1,1,1,0		29	32			26		22		14	62	
31		0,1,1,1,1,0	32			29		27		23		15	63	
32	•	1,1,1,1,1,0		31		30		28		24		16	64	
33	•	0,0,0,0,0,1	34		35		37		41		49			1
34		1,0,0,0,0,1		33	36		38		42		50			2
35		0,1,0,0,0,1	36			33	39		43		51			3
36		1,1,0,0,0,1		35		34	40		44		52			4
37		0,0,1,0,0,1	38		39			33	45		53			5
38		1,0,1,0,0,1		37	40			34	46		54			6
39		0,1,1,0,0,1	40			37		35	47		55			7
40		1,1,1,0,0,1		39		38		36	48		56			8
41		0,0,0,1,0,1	42		43		45			33	57			9
42		1,0,0,1,0,1		41	44		46			34	58			10
43		0,1,0,1,0,1	44			41	47			35	59			11
44		1,1,0,1,0,1		43		42	48			36	60			12
45		0,0,1,1,0,1	46		47			41		37	61			13
46		1,0,1,1,0,1		45	48			42		38	62			14
47		0,1,1,1,0,1	48			45		43		39	63			15
48		1,1,1,1,0,1		47		46		44		40	64			16
49		0,0,0,0,1,1	50		51		53		57			33		17
50		1,0,0,0,1,1		49	52		54		58			34		18
51		0,1,0,0,1,1	52			49	55		59			35		19
52		1,1,0,0,1,1		51		50	56		60			36		20
53		0,0,1,0,1,1	54		55			49	61			37		21
54		1,0,1,0,1,1		53	56			50	62			38		22
55		0,1,1,0,1,1	56			53		51	63			39		23
56		1,1,1,0,1,1		55		54		52	64			40		24
57		0,0,0,1,1,1	58		59		61			49		41		25
58		1,0,0,1,1,1		57	60		62			50		42		26
59		0,1,0,1,1,1	60			57	63			51		43		27
60		1,1,0,1,1,1		59		58	64			52		44		28
61		0,0,1,1,1,1	62		63			57		53		45		29
62		1,0,1,1,1,1		61	64			58		54		46		30
63		0,1,1,1,1,1	64			61		59		55		47		31
64		1,1,1,1,1,1		63		62		60		56		48		32

Table A.1: *a-machine* corresponding to the augmented burner model.

A.2 Dynamic Specifications

S	M	State variables						11	12	21	22	31	32	41	42	51	52	61	62
1	•	∞_1	∞_2	∞_3	∞_4	∞_5	∞_6	2	1	1	1	1	1	1	1	1	1	1	1
2	•	1	∞_2	∞_3	∞_4	∞_5	∞_6					3		1	1				
3		1	∞_2	1	∞_4	∞_5	∞_6					1		1	1				

Table A.2: *a–machine* corresponding to specification D1a.

S	M	State variables						11	12	21	22	31	32	41	42	51	52	61	62
1	•	∞_1^1	∞_2^1	∞_3^1	0	∞_5	∞_6	13		15		17		2		1	1	1	1
2	•	∞_1^1	∞_2^1	∞_3^1	∞_4^0	∞_5	∞_6	3		5		7			1	2	2	2	2
3		1	∞_2^1	∞_3^1	∞_4^0	∞_5	∞_6		2	4		8			13	3	3	3	3
4	•	1	1	∞_3^1	∞_4^0	∞_5	∞_6			5		3	9		14	4	4	4	4
5		∞_1^1	1	∞_3^1	∞_4^0	∞_5	∞_6	4				2	6		15	5	5	5	5
6		∞_1^1	1	1	∞_4^0	∞_5	∞_6	9				7		5	16	6	6	6	6
7		∞_1^1	∞_2^1	1	∞_4^0	∞_5	∞_6	8			6			2	17	7	7	7	7
8		1	∞_2^1	1	∞_4^0	∞_5	∞_6		7		9			3	18	8	8	8	8
9		1	1	1	∞_4^0	∞_5	∞_6	6			8			4	10	9	9	9	9
10		1	1	1	0	∞_5	∞_6					12		11			2		
11		1	1	1	1	∞_5	∞_6							4	10				
12		1	0	1	0	∞_5	∞_6						13	8					
13		1	∞_2^1	∞_3^1	0	∞_5	∞_6		1	14		18			3	13	13	13	13
14		1	1	∞_3^1	0	∞_5	∞_6		15			13	10		4	14	14	14	14
15		∞_1^1	1	∞_3^1	0	∞_5	∞_6	14				1	16		5	15	15	15	15
16		∞_1^1	1	1	0	∞_5	∞_6	10				17		15	6	16	16	16	16
17		∞_1^1	∞_2^1	1	0	∞_5	∞_6	18			16			1	7	17	17	17	17
18		1	∞_2^1	1	0	∞_5	∞_6		17		10			13	8	18	18	18	18

Table A.3: *a–machine* corresponding to specification D2a.

S	M	State variables						11	12	21	22	31	32	41	42	51	52	61	62
1	•	0	0	0	0	0	0	2						64					
2		∞_1^0	0	0	0	0	0		1	49		55		27		41		3	
3		∞_1^0	0	0	0	0	∞_6^0	60	48	18				26		4			2
4		∞_1^0	0	0	0	∞_5^0	∞_6^0	43	13	17				5			3		41
5		∞_1^0	0	0	∞_4^0	∞_5^0	∞_6^0	10	12	6					4		26		28
6		∞_1^0	0	∞_3^0	∞_4^0	∞_5^0	∞_6^0	9	7				5		17		19		29
7		∞_1^0	∞_2^0	∞_3^0	∞_4^0	∞_5^0	∞_6^0	8			6			12	14		20		30
8		0	∞_2^0	∞_3^0	∞_4^0	∞_5^0	∞_6^0	7			9			11	15		21		31
9		0	0	∞_3^0	∞_4^0	∞_5^0	∞_6^0	6		8				10	16		22		32
10		0	0	0	∞_4^0	∞_5^0	∞_6^0	5		11		9			43		23		33
11		0	∞_2^0	0	∞_4^0	∞_5^0	∞_6^0	12		10	8				44		24		34
12		∞_1^0	∞_2^0	0	∞_4^0	∞_5^0	∞_6^0	11		5	7				13		25		35
13		∞_1^0	∞_2^0	0	0	∞_5^0	∞_6^0	44			4	14		12			48		36
14		∞_1^0	∞_2^0	∞_3^0	0	∞_5^0	∞_6^0	15		17		13	7				47		37
15		0	∞_2^0	∞_3^0	0	∞_5^0	∞_6^0	14		16				44	8		46		38
16		0	0	∞_3^0	0	∞_5^0	∞_6^0	17	15					43	9		59		39
17		∞_1^0	0	∞_3^0	0	∞_5^0	∞_6^0	16	14					4	6		18		40
18		∞_1^0	0	∞_3^0	0	∞_5^0	∞_6^0	59	47					3	19	17			55
19		∞_1^0	0	∞_3^0	∞_4^0	0	∞_6^0	22	20					26		18	6		54
20		∞_1^0	∞_2^0	∞_3^0	∞_4^0	0	∞_6^0	21		19				25	47		7		51
21		0	∞_2^0	∞_3^0	∞_4^0	0	∞_6^0	20			22			24	46		8		52
22		0	0	∞_3^0	∞_4^0	0	∞_6^0	19		21				23	59		9		53
23		0	0	0	∞_4^0	0	∞_6^0	26	24			22			60		10		64
24		0	∞_2^0	0	∞_4^0	0	∞_6^0	25				23	21		45		11		63
25		∞_1^0	∞_2^0	0	∞_4^0	0	∞_6^0	24		26		20			48		12		50
26		∞_1^0	0	0	∞_4^0	0	∞_6^0	23	25			19			3		5		27
27		∞_1^0	0	0	∞_4^0	0	0	64	50			54		2	28			26	
28		∞_1^0	0	0	∞_4^0	∞_5^0	0	33	35			29			41		27	5	
29		∞_1^0	0	∞_3^0	∞_4^0	∞_5^0	0	32	30					28	40		54	6	
30		∞_1^0	∞_2^0	∞_3^0	∞_4^0	∞_5^0	0	31		29				35	37		51	7	
31		0	∞_2^0	∞_3^0	∞_4^0	∞_5^0	0	30				32		34	38		52	8	
32		0	0	∞_3^0	∞_4^0	∞_5^0	0	29		31				33	39		53	9	
33		0	0	0	∞_4^0	∞_5^0	0	28		34		32			42		64	10	
34		0	∞_2^0	0	∞_4^0	∞_5^0	0	35				33	31		62		63	11	
35		∞_1^0	∞_2^0	0	∞_4^0	∞_5^0	0	34		28		30			36		50	12	
36		∞_1^0	∞_2^0	0	0	∞_5^0	0	62		41	37			35			49	13	
37		∞_1^0	∞_2^0	∞_3^0	0	∞_5^0	0	38		40		36	30				56	14	
38		0	∞_2^0	∞_3^0	0	∞_5^0	0	37		39				62	31		57	15	
39		0	0	∞_3^0	0	∞_5^0	0	40	38					42	32		58	16	
40		∞_1^0	0	∞_3^0	0	∞_5^0	0	39	37					41	29		55	17	
41		∞_1^0	0	0	0	∞_5^0	0	42	36			40			28		2	4	
42	•	0	0	0	0	∞_5^0	0	41	62			39			33	1	43		
43		0	0	0	0	∞_5^0	∞_6^0	4	44			16		10			60		42
44		0	∞_2^0	0	0	∞_5^0	∞_6^0	13		43	15			11			45		62
45		0	∞_2^0	0	0	0	∞_6^0	48		60	46			24		44			61
46		0	∞_2^0	∞_3^0	0	0	∞_6^0	47		59		45	21		15				57
47		∞_1^0	∞_2^0	∞_3^0	0	0	∞_6^0	46	18			48	20		14				56
48		∞_1^0	∞_2^0	0	0	0	∞_6^0	45	3	47				25	13				49
49		∞_1^0	∞_2^0	0	0	0	0	61	2	56				50		36		48	
50	•	∞_1^0	∞_2^0	0	∞_4^0	0	0	63	27	51					49	35		25	
51		∞_1^0	∞_2^0	∞_3^0	∞_4^0	0	0	52	54			50			56	30		20	
52		0	∞_2^0	∞_3^0	∞_4^0	0	0	51	53			63			57	31		21	
53		0	0	∞_3^0	∞_4^0	0	0	54			52	64			58	32		22	
54		∞_1^0	0	∞_3^0	∞_4^0	0	0	53		51				27	55	29		19	
55		∞_1^0	0	∞_3^0	0	0	0	58	56					2	54	40		18	
56		∞_1^0	∞_2^0	∞_3^0	0	0	0	57		55		49	51		37			47	
57		0	∞_2^0	∞_3^0	0	0	0	56	58			61	52		38			46	
58		0	0	∞_3^0	0	0	0	55	57			1	53		39		59		
59		0	0	∞_3^0	0	0	∞_6^0	18	46			60	22		16				58
60	•	0	0	0	0	0	∞_6^0	3	45			59		23		43			1
61		0	∞_2^0	0	0	0	0	49		1	57			63	62			45	
62		0	∞_2^0	0	0	∞_5^0	0	36		42	38			34		61	44		
63		0	∞_2^0	0	∞_4^0	0	0	50				64	52		61	34		24	
64		0	0	0	∞_4^0	0	0	27		63		53		1	33		23		

Table A.4: *a–machine* corresponding to specification D1b.

S	M	State variables						TRANSITIONS											
								11	12	21	22	31	32	41	42	51	52	61	62
1	•	∞^1	0	∞_3^1	0	0	0	63		61		34		55		48		2	
2	•	∞^1	0	∞_3^1	0	0	∞^0	31		29		33		23		3			1
3		∞^1	0	∞_3^1	0	∞_5^0	∞_6^0	16		14		12		4			2		48
4		∞^1	0	∞_3^1	∞^0	∞_5^0	∞_6^0	7		5		11			3		23		37
5		∞^1	∞_2^0	∞_3^1	∞^0	∞_5^0	∞_6^0	6			4	10			14		22		38
6		1	∞_2^0	∞_3^1	∞_4^0	∞_5^0	∞_6^0		5		7	9			15		21		39
7		1	0	∞_3^1	∞_4^0	∞_5^0	∞_6^0		4	6		8			16		24		40
8		1	0	1	∞_4^0	∞_5^0	∞_6^0	11	9				7		17		25		41
9		1	∞_2^0	1	∞_4^0	∞_5^0	∞_6^0	10		8			6		18		20		42
10		∞^1	∞_2^0	1	∞_4^0	∞_5^0	∞_6^0	9		11			5		13		27		43
11		∞^1	0	1	∞_4^0	∞_5^0	∞_6^0	8		10			4		12		26		36
12		∞^1	0	1	0	∞_5^0	∞_6^0	17	13				3	11			33		35
13		∞^1	∞_2^0	1	0	∞_5^0	∞_6^0	18		12		14	10				28		44
14		∞^1	∞_2^0	∞_3^1	0	∞_5^0	∞_6^0	15			3	13		5			29		45
15		1	∞_2^0	∞_3^1	0	∞_5^0	∞_6^0		14	16	18		6				30		46
16		1	0	∞_3^1	0	∞_5^0	∞_6^0	3	15		17		7				31		47
17		1	0	1	0	∞_5^0	∞_6^0	12	18		16	8				.	32		49
18		1	∞_2^0	1	0	∞_5^0	∞_6^0	13		17		15	9				19		50
19		1	∞_2^0	1	0	0	∞_6^0	28		32		30	20		18				51
20		1	∞_2^0	1	∞_4^0	0	∞_6^0	27		25		21		19	9				52
21		1	∞_2^0	∞_3^1	∞_4^0	0	∞_6^0	22		24	20			30	6				53
22		∞^1	∞_2^0	∞_3^1	∞_4^0	0	∞_6^0	21		23	27			29	5				54
23		∞^1	0	∞_3^1	∞_4^0	0	∞_6^0	24		22		26		2	4				55
24		1	0	∞_3^1	∞_4^0	0	∞_6^0			23	21	25		31	7				56
25		1	0	1	∞_4^0	0	∞_6^0	26	20			24		32	8				57
26		∞^1	0	1	∞_4^0	0	∞_6^0	25		27		23		33	11				58
27		∞^1	∞_2^0	1	∞_4^0	0	∞_6^0	20			26	22		28	10				59
28		∞^1	∞_2^0	1	0	0	∞_6^0	19		33		29	27	13					60
29		∞^1	∞_2^0	∞_3^1	0	0	∞_6^0	30			2	28		22		14			61
30		1	∞_2^0	∞_3^1	0	0	∞_6^0	29		31	19		21		15				62
31		1	0	∞_3^1	0	0	∞_6^0	2	30			32		24		16			63
32		1	0	1	0	0	∞_6^0	33	19			31	25	17					64
33		∞^1	0	1	0	0	∞_6^0	32		28		2	26	12					34
34		∞^1	0	1	0	0	0	64		60		1	58		35			33	
35		∞^1	0	1	0	∞_5^0	0	49		44				48	36		34	12	
36		∞^1	0	1	∞_4^0	∞_5^0	0	41		43				37		35	58	11	
37		∞^1	0	∞_3^1	∞_4^0	∞_5^0	0	40		38			36		48		55	4	
38		∞^1	∞_2^0	∞_3^1	∞_4^0	∞_5^0	0	39				37	43		45		54	5	
39		1	∞_2^0	∞_3^1	∞_4^0	∞_5^0	0		38			40	42		46		53	6	
40		1	0	∞_3^1	∞_4^0	∞_5^0	0		37	39			41		47		56	7	
41		1	0	1	∞_4^0	∞_5^0	0	36	42					40	49		57	8	
42		1	∞_2^0	1	∞_4^0	∞_5^0	0	43				41		39	50		52	9	
43		∞^1	∞_2^0	1	∞_4^0	∞_5^0	0	42				36		38	44		59	10	
44		∞^1	∞_2^0	1	0	∞_5^0	0	50				35		45	43		60	13	
45		∞^1	∞_2^0	∞_3^1	0	∞_5^0	0	46				48	44		38		61	14	
46		1	∞_2^0	∞_3^1	0	∞_5^0	0		45			47	50		39		62	15	
47		1	0	∞_3^1	0	∞_5^0	0		48	46			49		40		63	16	
48	•	∞^1	0	∞_3^1	0	∞_5^0	0	47		45			35		37		1	3	
49		1	0	1	0	∞_5^0	0	35		50				47	41		64	17	
50		1	∞_2^0	1	0	∞_5^0	0	44		49				46	42		51	18	
51		1	∞_2^0	1	0	0	0	60		64				62	52	50		19	
52	•	1	∞_2^0	1	∞_4^0	0	0	59		57			53			51	42	20	
53	•	1	∞_2^0	∞_3^1	∞_4^0	0	0	54		56	52				62	39		21	
54		∞^1	∞_2^0	∞_3^1	∞_4^0	0	0	53			55	59			61	38		22	
55		∞^1	0	∞_3^1	∞_4^0	0	0	56		54			58			1	37	23	
56		1	0	∞_3^1	∞_4^0	0	0	55	53				57			63	40	24	
57		1	0	1	∞_4^0	0	0	58	52				56			64	41	25	
58		∞^1	0	1	∞_4^0	0	0	57			59		55			34	36	26	
59		∞^1	∞_2^0	1	∞_4^0	0	0	52			58		54			60	43	27	
60		∞^1	∞_2^0	1	0	0	0	51		34			61	59			44	28	
61		∞^1	∞_2^0	∞_3^1	0	0	0	62			1	60		54			45	29	
62		1	∞_2^0	∞_3^1	0	0	0		61		63	51		53			46	30	
63		1	0	∞_3^1	0	0	0	1	62			64		56			47	31	
64		1	0	1	0	0	0	34	65				63	57			49	32	
65		1	1	1	0	0	0						64		52				

Table A.5: *a-machine* corresponding to specification D2b.

S	M	State variables						TRANSITIONS											
								11	12	21	22	31	32	41	42	51	52	61	62
1	•	∞	∞	0	∞	0	0	60	54		56		66		45		2		
2	•	∞	∞	0	0	0	∞	27	25	21	19						3		1
3		∞	∞	0	∞	∞	∞	10	8		4				18		2		45
4		∞	∞	∞	∞	∞	∞	5	7					3	13		21		40
5		1	∞	∞	∞	∞	∞		4	6				10	12		22		39
6		1	1	∞	∞	∞	∞		7				5	9	15		23		42
7		∞	1	∞	∞	∞	∞	6			4			8	14		24		41
8		∞	1	0	∞	∞	∞	9				3	7		17		25		44
9		1	1	0	∞	∞	∞		8			10	6		16		26		43
10		1	∞	0	∞	∞	∞		3	9			5		11		27		38
11		1	∞	0	1	∞	∞	18	16				12		10		28		37
12		1	∞	∞	1	∞	∞	13	15				11		5		29		48
13		∞	∞	∞	1	∞	∞	12	14				18		4		20		47
14		∞	1	∞	1	∞	∞	15			13		17		7		31		50
15		1	1	∞	1	∞	∞		14		12		16		6		30		49
16		1	1	0	1	∞	∞		17		11	15			9		33		52
17		∞	1	0	∞	∞	∞	16			18	14			8		32		51
18		∞	∞	0	1	∞	∞	11		17		13			3		19		46
19		∞	∞	0	1	0	∞	28		32		20			2	18			66
20		∞	∞	∞	1	0	∞	29		31				19		21	13		63
21		∞	∞	∞	1	0	∞	22		24				2	20		4		56
22		1	∞	∞	1	0	∞		21	23				27	29		5		57
23		1	1	∞	1	0	∞		24		22			26	30		6		58
24		∞	1	∞	1	0	∞		23		21			25	31		7		55
25		∞	1	0	1	0	∞		26			2	24		32		8		54
26		1	1	0	1	0	∞		25			27	23		33		9		59
27		1	∞	0	1	0	∞		2	26		22			28		10		60
28		1	∞	0	1	0	∞	19		33		29				27	11		61
29		1	∞	∞	1	0	∞	20		30				28		22	12		62
30		1	1	∞	1	0	∞	31		29				33		23	15		65
31		∞	1	∞	1	0	∞	30		20				32		24	14		64
32		∞	1	0	1	0	∞	33		19	31				25	17			53
33		1	1	0	1	0	∞	32		28	30				26	16			34
34	•	1	1	0	1	0	0		35						59				
35		1	0	0	1	0	0								36	37			
36		1	0	0	0	0	0	1						35					
37		1	∞	0	1	∞	0	46	52		48				38			61	11
38		1	∞	0	∞	∞	0	45	43		39			37				60	10
39		1	∞	∞	∞	∞	0	40	42					38	48			57	5
40		∞	∞	∞	∞	∞	0	39	41				45	47			56	4
41		∞	1	∞	∞	∞	0	42		40			44	50			55	7
42		1	1	∞	1	∞	0	41		39			43	49			58	6
43		1	1	0	1	∞	0	44		38	42			52			59	9
44	•	∞	1	0	1	∞	0	43		45	41			51			54	8
45		∞	∞	0	1	∞	0	38		44		40			46		1	3
46		∞	∞	0	1	∞	0	37		51		47			45		66	18
47		∞	∞	∞	1	∞	0	48		50			46	40			63	13
48		1	∞	∞	1	∞	0	47	49				37	39			62	12
49		1	1	∞	1	∞	0	50		48			52	42			65	15
50		∞	1	∞	1	∞	0	49		47			51	41			64	14
51		∞	1	0	1	∞	0	52		46	50			44			53	17
52		1	1	0	1	∞	0	51		37	49			43			34	16
53		∞	1	0	1	0	0	34		66	64			54	51			32
54		∞	1	0	∞	0	0	59		1	55		53		44			25
55		∞	1	∞	∞	0	0	58		56		54	64		41			24
56		∞	∞	∞	∞	0	0	57		55		1	63		40			21
57		1	∞	∞	∞	0	0	56	58		57		60	62			39	22
58		1	1	∞	∞	0	0	55		57		59	65		42			23
59		1	1	0	∞	0	0	54		60	58			34			43	26
60		1	∞	0	∞	0	0	1	59		57			61			38	27
61		1	∞	0	1	0	0	66	34		62			60	37			28
62		1	∞	∞	1	0	0	63	65				61	57	48			29
63		∞	∞	∞	1	0	0	62		64			66	56	47			20
64		∞	1	∞	1	0	0	65		63			53	55	50			31
65		1	1	∞	1	0	0	64		62			34	58	49			30
66		∞	∞	0	1	0	0	61	53		63			1	46			19

Table A.6: *a-machine* corresponding to specification D3.

S	M	State variables	11	12	21	22	31	32	41	42	51	52	61	62
1	•	$\infty_1\ \ \infty_2\ \ \infty_3\ \ \infty_4\ \ \infty_5\ \ \infty_6$	1	1	1	1	1	2	1	1	1	1	1	1
2	•	$\infty_1\ \ \infty_2\ \ 0\ \ \infty_4\ \ \infty_5\ \ \infty_6$	1	1					1	1				

Table A.7: *a–machine* corresponding to specification D4.

S	M	State variables	11	12	21	22	31	32	41	42	51	52	61	62
1	•	$\infty_1\ \ 0\ \ \infty_3\ \ 0\ \ 0\ \ 0$	1	1	15		1	1	17		14		2	
2	•	$\infty_1\ \ 0\ \ \infty_3\ \ 0\ \ 0\ \ \infty_6^{0}$	2	2	7		2	2	9		3			1
3		$\infty_1\ \ 0\ \ \infty_3\ \ 0\ \ \infty_5^{0}\ \ \infty_6^{0}$	3	3	6		3	3	4			2		14
4		$\infty_1\ \ 0\ \ \infty_3\ \ \infty_4^{0}\ \ \infty_5^{0}\ \ \infty_6^{0}$	4	4	5		4	4		3		9		11
5		$\infty_1\ \ \infty_2^{0}\ \ \infty_3\ \ \infty_4^{0}\ \ \infty_5^{0}\ \ \infty_6^{0}$	5	5	5	4	5	5		6		8		12
6		$\infty_1\ \ \infty_2^{0}\ \ \infty_3\ \ 0\ \ \infty_5^{0}\ \ \infty_6^{0}$	6	6		3	6	6	5			7		13
7		$\infty_1\ \ \infty_2^{0}\ \ \infty_3\ \ 0\ \ 0\ \ \infty_6^{0}$	7	7		2	7	7	8		6			15
8		$\infty_1\ \ \infty_2^{0}\ \ \infty_3\ \ \infty_4^{0}\ \ 0\ \ \infty_6^{0}$	8	8		9	8	8		7		5		16
9		$\infty_1\ \ 0\ \ \infty_3\ \ \infty_4^{0}\ \ 0\ \ \infty_6^{0}$	9	9	8		9	9		2		4		10
10		$\infty_1\ \ 0\ \ \infty_3\ \ \infty_4^{0}\ \ 0\ \ 0$	10	10	16		10	10		1	11		9	
11		$\infty_1\ \ 0\ \ \infty_3\ \ \infty_4^{0}\ \ \infty_5^{0}\ \ 0$	11	11	12		11	11		14		10	4	
12		$\infty_1\ \ \infty_2^{0}\ \ \infty_3\ \ \infty_4^{0}\ \ \infty_5^{0}\ \ 0$	12	12		11	12	12		13		16	5	
13		$\infty_1\ \ \infty_2^{0}\ \ \infty_3\ \ 0\ \ \infty_5^{0}\ \ 0$	13	13		14	13	13	12			15	6	
14	•	$\infty_1\ \ 0\ \ \infty_3\ \ 0\ \ \infty_5^{0}\ \ 0$	14	14	13		14	14	11			1	3	
15		$\infty_1\ \ \infty_2^{0}\ \ \infty_3\ \ 0\ \ 0\ \ 0$	15	15		1	15	15	16		13		7	
16	•	$\infty_1\ \ \infty_2^{0}\ \ \infty_3\ \ \infty_4^{0}\ \ 0\ \ 0$	16	16		10	16	16		15		12	8	
17		$\infty_1\ \ 0\ \ \infty_3\ \ 1\ \ 0\ \ 0$										11		

Table A.8: *a–machine* corresponding to specification D5a.

S	M	State variables						11	12	21	22	31	32	41	42	51	52	61	62
1	•	∞^1_2	∞^1_2	0	∞^1_1	0	0	56		54		52		64		42		2	
2	•	∞^1_2	∞^2_4	0	∞^4_4	0	∞^0_6	27		25		21		19		3			1
3		∞^1_2	∞^2_2	0	∞^1_4	∞^0_5	∞^0_6	10		8		4		18			2		42
4		∞^1_1	∞^1_2	∞^0_3	∞^1_4	∞^0_5	∞^0_6	5		7				3	13		21		39
5	1		∞^2_2	∞^3_3	∞^1_4	∞^0_5	∞^0_6			4	6			10	12		22		38
6	1	1	∞^3_3	∞^1_4	∞^0_5	∞^0_6				7		5		9	15		23		37
7		∞^1_1	1	∞^3_3	∞^1_4	∞^0_5	∞^0_6	6				4		8	14		24		40
8		∞^1_1	1	0	∞^1_4	∞^0_5	∞^0_6	9				3	7		17		25		41
9		1	∞^1_2	0	∞^1_4	∞^0_5	∞^0_6			8		10	6		16		26		36
10		1	∞^1_2	0	∞^1_4	∞^0_5	∞^0_6			3	9	5			11		27		43
11		1	∞^1_2	0	1	∞^0_5	∞^0_6	18	16			12			10		28		44
12		1	∞^2_2	∞^0_3	1	∞^0_5	∞^0_6	13	15					11		5	29		45
13		∞^1_1	∞^2_2	∞^3_3	1	∞^0_5	∞^0_6	12		14				18		4	20		46
14		∞^1_1	1	∞^3_3	1	∞^0_5	∞^0_6	15				13		17		7	31		47
15		1	1	∞^3_3	1	∞^0_5	∞^0_6			14		12		16		6	30		48
16		1	1	0	1	∞^0_5	∞^0_6			17		11	15			9	33		35
17		∞^1_1	1	0	1	∞^0_5	∞^0_6	16				18	14			8	32		62
18		∞^1_1	∞^1_2	0	1	∞^0_5	∞^0_6	11		17				13		3	19		63
19		∞^1_1	∞^1_2	0	1	0	∞^0_6	28		32		20				2	18		64
20		∞^1_1	∞^1_2	∞^0_3	1	0	∞^0_6	29		31				19		21	13		59
21		∞^1_1	∞^2_2	∞^3_3	∞^1_4	0	∞^0_6	22		24				2	20		4		52
22		1	∞^2_2	∞^3_3	∞^1_4	0	∞^0_6			21	23			27	29		5		51
23		1	1	∞^3_3	∞^1_4	0	∞^0_6			24		22		26	30		6		50
24		∞^1_1	1	∞^3_3	∞^1_4	0	∞^0_6	23				21		25	31		7		53
25		∞^1_1	1	0	∞^1_4	0	∞^0_6	26				2	24		32		8		54
26		1	1	0	∞^1_4	0	∞^0_6			25		27	23		33		9		55
27		1	∞^1_2	0	∞^1_4	0	∞^0_6			2	26			22		28	10		56
28		1	∞^1_2	0	1	0	∞^0_6	19		33		29				27	11		57
29		1	∞^1_2	∞^0_3	1	0	∞^0_6	20		30				28		22	12		58
30		1	1	∞^3_3	1	0	∞^0_6	31				29		33		23	15		49
31		∞^1_1	1	∞^3_3	1	0	∞^0_6	30				20		32		24	14		60
32		∞^1_1	1	0	1	0	∞^0_6	33				19	31			25	17		61
33		1	1	0	1	0	∞^0_6			32		28	30			26	16		34
34	•	1	1	0	1	0	0	61		65		49				55	35		33
35		1	1	0	1	∞^0_5	0	62		44		48				36		34	16
36		1	1	0	∞^1_4	∞^0_5	0	41		43		37			35		55		9
37		1	1	∞^0_3	∞^1_4	∞^0_5	0	40		38				36	48		50		6
38		1	∞^1_2	∞^3_3	∞^1_4	∞^0_5	0	39		37				43	45		51		5
39		∞^1_1	∞^2_2	∞^3_3	∞^1_4	∞^0_5	0	38		40				42	46		52		4
40		∞^1_1	1	∞^3_3	∞^1_4	∞^0_5	0	37				39		41	47		53		7
41		∞^1_1	1	0	∞^1_4	∞^0_5	0	36				42	40		62		54		8
42	•	∞^1_1	∞^1_2	0	∞^1_4	∞^0_5	0	43		41		39			63		1		3
43		1	∞^1_2	0	∞^1_4	∞^0_5	0	42		36		38			44		56		10
44		1	∞^1_2	0	1	∞^0_5	0	63		35		45				43	57		11
45		1	∞^2_2	∞^0_3	1	∞^0_5	0	46		48				44		38	58		12
46		∞^1_1	∞^2_2	∞^3_3	1	∞^0_5	0	45		47					63	39	59		13
47		∞^1_1	1	∞^3_3	1	∞^0_5	0	48				46			62	40	60		14
48		1	1	∞^3_3	1	∞^0_5	0	47				45			35	37	49		15
49		1	1	∞^3_3	1	0	0	60				58			34	50	48		30
50		1	1	∞^3_3	∞^1_4	0	0	53		51				55	49		37		23
51		1	∞^1_2	∞^3_3	∞^1_4	0	0	52		50				56	58		38		22
52		∞^1_1	∞^2_2	∞^3_3	∞^1_4	0	0	51		53				1	59		39		21
53		∞^1_1	1	∞^3_3	∞^1_4	0	0	50				52		54	60		40		24
54		∞^1_1	1	0	∞^1_4	0	0	55				1	53		61		41		25
55		1	1	0	∞^1_4	0	0	54		56		50			34		36		26
56		1	∞^1_2	0	∞^1_4	0	0	1		55		51			57		43		27
57		1	∞^1_2	0	1	0	0	64		34		58				56	44		28
58		1	∞^1_2	∞^0_3	1	0	0	59		49					57	51	45		29
59		∞^1_1	∞^2_2	∞^0_3	1	0	0	58		60					64	52	46		20
60		∞^1_1	1	∞^0_3	1	0	0	49		59					61	53	47		31
61		∞^1_1	1	0	1	0	0	34				64	60			54	62		32
62		∞^1_1	1	0	1	∞^0_5	0	35				63	47		41			61	17
63		∞^1_1	∞^1_2	0	1	∞^0_5	0	44		62		46			42			64	18
64		∞^1_1	∞^1_2	0	1	0	0	57		61		59			1	63			19
65		1	0	0	1	0	0									56	44		

Table A.9: *a-machine* corresponding to specification D5b.

S	M	State variables						11	12	21	22	31	32	41	42	51	52	61	62
								TRANSITIONS											
1	•	∞_1^1	∞_2^1	∞_3	∞_4^1	0	0	30	18			1	1	28		20		2	
2	•	∞_1^1	∞_2^1	∞_3	∞_4^1	0	∞_6^0	13	17			2	2	15		3			1
3		∞_1^1	∞_2^1	∞_3	∞_4^1	∞_5^0	∞_6^0	4	6			3	3	8			2		20
4		1	∞_2^1	∞_3	∞_4^1	∞_5^0	∞_6^0		3		5	4	4	9			13		21
5		1	1	∞_3	∞_4^1	∞_5^0	∞_6^0		6	4		5	5	10			12		22
6		∞_1^1	1	∞_3	∞_4^1	∞_5^0	∞_6^0	5		3		6	6	7			17		19
7		∞_1^1	1	∞_3	1	∞_5^0	∞_6^0	10		8		7	7		6		16		26
8		∞_1^1	∞_2^1	∞_3	1	∞_5^0	∞_6^0	9		7		8	8		3		15		25
9		1	∞_2^1	∞_3	1	∞_5^0	∞_6^0			8	10	9	9		4		14		24
10		1	1	∞_3	1	∞_5^0	∞_6^0			7	9	10	10		5		11		23
11		1	1	∞_3	1	0	∞_6^0	16	14			11	11	12	10				32
12		1	1	∞_3	∞_4^1	0	∞_6^0	17	13			12	12	11		5			31
13		1	∞_2^1	∞_3	∞_4^1	0	∞_6^0	2	12			13	13	14			4		30
14		1	∞_2^1	∞_3	1	0	∞_6^0	15	11			14	14		13		9		29
15		∞_1^1	∞_2^1	∞_3	1	0	∞_6^0	14	16			15	15		2		8		28
16		∞_1^1	1	∞_3	1	0	∞_6^0	11		15		16	16		17		7		27
17		∞_1^1	1	∞_3	∞_4^1	0	∞_6^0	12		2		17	17	16			6		18
18		∞_1^1	1	∞_3	∞_4^1	0	0	31		1		18	18	27		19		17	
19		∞_1^1	1	∞_3	∞_4^1	∞_5^0	0	22		20		19	19	26			18	6	
20	•	∞_1^1	∞_2^1	∞_3	∞_4^1	∞_5^0	0	21		19		20	20	25			1	3	
21		1	∞_2^1	∞_3	∞_4^1	∞_5^0	0	20	22			21	21	24			30	4	
22		1	1	∞_3	∞_4^1	∞_5^0	0	19		21		22	22	23			31	5	
23		1	1	∞_3	1	∞_5^0	0	26	24			23	23		22		32		10
24		1	∞_2^1	∞_3	1	∞_5^0	0	25	23			24	24		21		29		9
25		∞_1^1	∞_2^1	∞_3	1	∞_5^0	0	24	26			25	25		20		28		8
26		∞_1^1	1	∞_3	1	∞_5^0	0	23		25		26	26		19		27		7
27		∞_1^1	1	∞_3	1	0	0	32		28		27	27		18	26			16
28		∞_1^1	∞_2^1	∞_3	1	0	0	29		27		28	28		1		25		15
29		1	∞_2^1	∞_3	1	0	0	28	32			29	29		30		24		14
30		1	∞_2^1	∞_3	∞_4^1	0	0	1	31			30	30	29			21		13
31		1	1	∞_3	∞_4^1	0	0	18		30		31	31	32			22		12
32	•	1	1	∞_3	1	0	0	27		29		32	32		33		23		11
33		1	1	∞_3	0	0	0												12

Table A.10: *a-machine* corresponding to specification D6.

S	M	State variables						11	12	21	22	31	32	41	42	51	52	61	62
								TRANSITIONS											
1	•	∞_1	∞_2	∞_3	0	∞_5	∞_6	1	1	1	1	1	1	2		1		1	1
2	•	∞_1	∞_2	∞_3	∞_4^0	∞_5	∞_6	2	2	2	2	2	2		1	2	2	2	2

Table A.11: *a-machine* corresponding to specification D7.

S	M	State variables						11	12	21	22	31	32	41	42	51	52	61	62
						TRANSITIONS													
1	•	∞_1	∞_2	∞_3	∞_4^1	∞_5	∞_6	1	1	1	1	1	1	2		1	1	1	1
2	•	∞_1	∞_2	∞_3	1	∞_5	∞_6	2	2	2	2	2	2		1	2	2		2

Table A.12: *a-machine* corresponding to specification D8a.

S	M	State variables						11	12	21	22	31	32	41	42	51	52	61	62
						TRANSITIONS													
1	•	∞_1	0	∞_3	0	∞_5	∞_6	1	1	4		1	1	2		1	1		1
2		∞_1	0	∞_3	∞_4^0	∞_5	∞_6	2	2	3		2	2	2	1	2	2	2	2
3	•	∞_1	∞_2^0	∞_3	∞_4^0	∞_5	∞_6	3	3	3	2	3	3	3	4	3	3	3	3
4		∞_1	∞_2^0	∞_3	0	∞_5	∞_6	4	4	4	1	4	4	3		4	4	4	4

Table A.13: *a-machine* corresponding to specification D8b.

S	M	State variables						11	12	21	22	31	32	41	42	51	52	61	62
						TRANSITIONS													
1	•	∞_1	∞_2	∞_3	∞_4	∞_5^1	∞_6	1	1	1	1	1	1	1	1	2		1	1
2	•	∞_1	∞_2	∞_3	∞_4	1	∞_6	2		2		2	2	2					

Table A.14: *a-machine* corresponding to specification D9a.

S	M	State variables						11	12	21	22	31	32	41	42	51	52	61	62
						TRANSITIONS													
1	•	∞_1	∞_2	∞_3	∞_4	∞_5	∞_6^1	1	1	1	1	1	1	1	1	1	1	2	
2	•	∞_1	∞_2	∞_3	∞_4	∞_5	1	2		2		2	2	2					

Table A.15: *a-machine* corresponding to specification D9b.

S	M	State variables						11	12	21	22	31	32	41	42	51	52	61	62
1	•	∞_1	0	∞_3^1	∞_4^1	∞_5^1	0	1	1	20		22		24		26		2	
2	•	∞_1	0	∞_3^1	∞_4^1	∞_5^1	∞_6^0	2	2	3		5		7		13			1
3		∞_1	∞_2^0	∞_3^1	∞_4^1	∞_5^1	∞_6^0	3	3		2	4		8		12			20
4		∞_1	∞_2^0	1	∞_4^1	∞_5^1	∞_6^0	4	4			5	3	9		11			21
5		∞_1	0	1	∞_4^1	∞_5^1	∞_6^0	5	5	4			2	6		14			22
6		∞_1	0	1	1	∞_5^1	∞_6^0	6	6	9			7		5	15			23
7		∞_1	0	∞_3^1	1	∞_5^1	∞_6^0	7	7	8		6			2	16			24
8		∞_1	∞_2^0	∞_3^1	1	∞_5^1	∞_6^0	8	8		7	9			3	17			19
9		∞_1	∞_2^0	1	1	∞_5^1	∞_6^0	9	9		6	8			4	10			32
10		∞_1	∞_2^0	1	1	1	∞_6^1	10	10	15		17		11			9		31
11		∞_1	∞_2^0	1	∞_4^1	1	∞_6^0	11	11	14		12	10				4		28
12		∞_1	∞_2^0	∞_3^1	∞_4^1	1	∞_6^0	12	12	13		11		17			3		27
13		∞_1	0	∞_3^1	∞_4^1	1	∞_6^0	13	13	12		14		16			2		26
14		∞_1	0	1	∞_4^1	1	∞_6^0	14	14	11			13	15			5		29
15		∞_1	0	1	1	1	∞_6^0	15	15	10			16		14		6		30
16		∞_1	0	∞_3^1	1	1	∞_6^0	16	16	17		15			13		7		25
17		∞_1	∞_2^0	∞_3^1	1	1	∞_6^0	17	17	16		10			12		8		18
18		∞_1	∞_2^0	∞_3^1	1	1	0	18	18	25	31				27		19	17
19	•	∞_1	∞_2^0	∞_3^1	1	∞_5^1	0	19	19	24	32				20	18		8
20		∞_1	∞_2^0	∞_3^1	∞_4^1	∞_5^1	0	20	20	1	21		19		27			3
21		∞_1	∞_2^0	1	∞_4^1	∞_5^1	0	21	21		22		20	32	28			4
22		∞_1	0	1	∞_4^1	∞_5^1	0	22	22	21			1	23	29			5
23		∞_1	0	1	1	∞_5^1	0	23	23	32			24		22	30		6
24		∞_1	0	∞_3^1	1	∞_5^1	0	24	24	19		23			1	25		7
25		∞_1	0	∞_3^1	1	1	0	25	25	18		30			26		24	16
26	•	∞_1	0	∞_3^1	∞_4^1	1	0	26	26	27		29		25		1		13
27		∞_1	∞_2^0	∞_3^1	∞_4^1	1	0	27	27		26	28		18			20	12
28		∞_1	∞_2^0	1	∞_4^1	1	0	28	28	29			27	31			21	11
29		∞_1	0	1	∞_4^1	1	0	29	29	28			26	30			22	14
30		∞_1	0	1	1	1	0							25			
31		∞_1	∞_2^0	1	1	1	0	31	31	30		18			28		32	10
32		∞_1	∞_2^0	1	1	∞_5^1	0	32	32	23		19		21	31			9

Table A.16: *a-machine* corresponding to specification D10a.

S	M	State variables						TRANSITIONS											
								11	12	21	22	31	32	41	42	51	52	61	62
1	•	∞^1	0	0	∞^1_4	∞^1_5	0	42		44		38		36		58		2	
2	•	∞^1_1	0	0	∞^1_4	∞^1_5	∞^0_6	7		9		3		15		25			1
3		∞^1_1	0	∞^0_3	∞^1_4	∞^1_5	∞^0_6	6		4			2	14		22			38
4		∞^0_1	∞^0_2	∞^0_3	∞^1_4	∞^1_5	∞^0_6	5				3		9	11	21			39
5		1	∞^0_2	∞^0_3	∞^1_4	∞^1_5	∞^0_6		4		6			8	12	20			40
6		1	0	∞^0_3	∞^1_4	∞^1_5	∞^0_6	3	5					7	13	23			41
7		1	0	0	∞^1_4	∞^1_5	∞^0_6	2	8			6		16		24			42
8		1	∞^0_2	0	∞^1_4	∞^1_5	∞^0_6		9			7	5	17		19			43
9		∞^1_1	∞^0_2	0	∞^1_4	1	∞^0_6	8				2	4	10		26			44
10		∞^1_1	∞^0_2	0	1	∞^1_5	∞^0_6	17				15	11		9	27			45
11		∞^1_1	∞^0_2	∞^0_3	1	∞^1_5	∞^0_6	12				14		10	4	28			46
12		1	∞^0_2	∞^0_3	1	∞^1_5	∞^0_6		11			13		17	5	29			47
13		1	0	∞^0_3	1	∞^1_5	∞^0_6		14	12				16	6	30			48
14		∞^1_1	0	∞^0_3	1	∞^1_5	∞^0_6	13		11				15	3	31			37
15		∞^1_1	0	0	1	∞^1_5	∞^0_6	16		10			14		2	32			36
16		1	0	0	1	∞^1_5	∞^0_6		15	17		13			7	33			49
17		1	∞^0_2	0	1	∞^1_5	∞^0_6		10			16	12		8	18			50
18		1	∞^0_2	0	1	1	∞^0_6		27			33	29		19		17		51
19		1	∞^0_2	0	∞^1_4	1	∞^0_6		26	24	20			18			8		52
20		1	∞^0_2	∞^0_3	∞^1_4	1	∞^0_6		21	23		19	29				5		53
21		∞^1_1	∞^0_2	∞^0_3	∞^1_4	1	∞^0_6	20				22		26	28		4		54
22		∞^1_1	0	∞^0_3	∞^1_4	1	∞^0_6	23		21				25	31		3		55
23		1	0	∞^0_3	∞^1_4	1	∞^0_6		22	20				24	30		6		56
24		1	0	0	∞^1_4	1	∞^0_6		25	19		23			33		7		57
25		∞^1_1	0	0	∞^1_4	1	∞^0_6	24		26		22			32		2		58
26		∞^1_1	∞^0_2	0	∞^1_4	1	∞^0_6	19				25	21		27		9		59
27		∞^1_1	∞^0_2	0	1	1	∞^0_6	18				32	28		26		10		60
28		∞^1_1	∞^0_2	∞^0_3	1	1	∞^0_6	29				31		27	21		11		61
29		1	∞^0_2	∞^0_3	1	1	∞^0_6		28	30				18	20		12		62
30		1	0	∞^0_3	1	1	∞^0_6		31	29				33	23		13		63
31		∞^1_1	0	∞^0_3	1	1	∞^0_6	30		28				32	22		14		64
32		∞^1_1	0	0	1	1	∞^0_6	33		27		31			25		15		35
33		1	0	0	1	1	∞^1_6		32	18		30			24		16		34
34		1	0	0	1	1	0		35										
35		∞^1_1	0	0	1	1	0	34		60		64			58		36	32	
36		∞^1_1	0	0	1	∞^1_5	0	49		45		37			1	35		15	
37		∞^1_1	0	∞^0_3	1	∞^1_5	0	48		46			36		38	64		14	
38		∞^1_1	0	∞^0_3	∞^1_4	∞^1_5	0	41		39			1	37		55		3	
39		∞^0_1	∞^0_2	∞^0_3	∞^1_4	∞^1_5	0	40				38		44	46	54		4	
40		1	∞^0_2	∞^0_3	∞^1_4	∞^1_5	0		39			41		43	47	53		5	
41		1	0	∞^0_3	∞^1_4	∞^1_5	0		38	40				42	48	56		6	
42		1	0	0	∞^1_4	∞^1_5	0		1	43		41		49		57		7	
43		1	∞^0_2	0	∞^1_4	∞^1_5	0		44			42	40	50		52		8	
44		∞^1_1	∞^0_2	0	∞^1_4	∞^1_5	0	43				1	39	45		59		9	
45		∞^1_1	∞^0_2	0	1	∞^1_5	0	50				36	46		44	60		10	
46		∞^1_1	∞^0_2	∞^0_3	1	∞^1_5	0	47				37		45	39	61		11	
47		1	∞^0_2	∞^0_3	1	∞^1_5	0		46			48		50	40	62		12	
48		1	0	∞^0_3	1	∞^1_5	0		37	47				49	41	63		13	
49		1	0	0	1	∞^1_5	0		36	50		48			42	34		16	
50	•	1	∞^0_2	0	1	∞^1_5	0		45			49	47		43	51		17	
51		1	∞^0_2	0	1	1	0		60	34	62				52		50	18	
52		1	∞^0_2	0	∞^1_4	1	0		59	57	53			51			43	19	
53		1	∞^0_2	∞^0_3	∞^1_4	1	0		54	56				52	62		40	20	
54		∞^1_1	∞^0_2	∞^0_3	∞^1_4	1	0	53				55		59	61		39	21	
55		∞^1_1	0	∞^0_3	∞^1_4	1	0	56		54				58	64		38	22	
56		1	0	∞^0_3	∞^1_4	1	0		55	53				57	63		41	23	
57		1	0	0	∞^1_4	1	0		58	52		56			34		42	24	
58	•	∞^1_1	0	0	∞^1_4	1	0	57		59		55			35		1	25	
59		∞^1_1	∞^0_2	0	∞^1_4	1	0	52				58	54		60		44	26	
60		∞^1_1	∞^0_2	0	1	1	0	51				35	61		59		45	27	
61		∞^1_1	∞^0_2	∞^0_3	1	1	0	62				64		60	54		46	28	
62		1	∞^0_2	∞^0_3	1	1	0		61	63				51	53		47	29	
63		1	0	∞^0_3	1	1	0		64	62				34	56		48	30	
64		∞^1_1	0	∞^0_3	1	1	0	63		61				35	55		37	31	

Table A.17: *a-machine* corresponding to specification D10b.

S	M	State variables						TRANSITIONS											
								11	12	21	22	31	32	41	42	51	52	61	62
1	•	∞_1	$\infty_2^{1/2}$	∞_3	∞_4	0	$\infty_6^{1/6}$	1	1	4		1	1	1	1	2		6	
2	• •	∞_1	$\infty_2^{1/2}$	∞_3	∞_4	$\infty_5^{0/8}$	$\infty_6^{1/6}$	2	2	3		2	2	2	2		1	7	
3		∞_1	1	∞_3	∞_4	$\infty_5^{0/5}$	$\infty_6^{1/6}$	3	3		2	3	3	3	3		4	8	
4	•	∞_1	1	∞_3	∞_4	0	∞_6^{1}	4	4	1		4	4	4	4	3		5	
5		∞_1	1	∞_3	∞_4	0	1				6								
6	•	∞_1	$\infty_2^{1/1}$	∞_3	∞_4	0	1	6	6	5		6	6	6	6	7			1
7		∞_1	$\infty_2^{1/2}$	∞_3	∞_4	$\infty_5^{0/0}$	1	7	7	8		7	7	7	7		6		2
8		∞_1	1	∞_3	∞_4	$\infty_5^{0/5}$	1	8	8		7	8	8	8	8		5		3

Table A.18: *a-machine* corresponding to specification D11a.

S	M	State variables						TRANSITIONS											
								11	12	21	22	31	32	41	42	51	52	61	62
1	•	∞_1	0	$\infty_3^{1/3}$	∞_4	0	$\infty_6^{1/6}$	1	1	8		6	1	1	1	2		16	1
2	•	∞_1	0	$\infty_3^{1/3}$	∞_4	$\infty_5^{0/8}$	$\infty_6^{1/6}$	2	2	3			5	2	2	2	1	11	2
3		∞_1	$\infty_2^{0/8}$	$\infty_3^{1/3}$	∞_4	$\infty_5^{0/8}$	$\infty_6^{1/6}$	3	3	3	2	4	3	3	3	3	8	10	3
4		∞_1	$\infty_2^{0/2}$	1	∞_4	$\infty_5^{1/5}$	$\infty_6^{1/6}$	4	4	4	5		3	4	4	4	7	13	4
5		∞_1	0	1	∞_4	$\infty_5^{0/5}$	$\infty_6^{1/6}$	5	5	4			2	5	5	5	6	12	5
6		∞_1	0	1	∞_4	0	$\infty_6^{1/6}$	6	6	7			1	6	6	5		15	6
7		∞_1	$\infty_2^{0/0}$	1	∞_4	0	$\infty_6^{1/6}$	7	7	7	6		8	7	7	4		14	7
8	•	∞_1	$\infty_2^{0/8}$	$\infty_3^{1/3}$	∞_4	0	$\infty_6^{1/6}$	8	8	8	1	7	8	8	8	3		9	8
9		∞_1	$\infty_2^{1/2}$	$\infty_3^{1/3}$	∞_4	0	1	9	9	9	16	14	9	9	9	10			8
10		∞_1	$\infty_2^{0/2}$	$\infty_3^{1/3}$	∞_4	$\infty_5^{0/8}$	1	10	10	10	11	13	10	10	10	10	9		3
11		∞_1	0	$\infty_3^{1/3}$	∞_4	$\infty_5^{0/8}$	1	11	11	10		12	11	11	11	11	16		2
12		∞_1	0	1	∞_4	$\infty_5^{0/8}$	1	12	12	13			11	12	12	12	15		5
13		∞_1	$\infty_2^{0/2}$	1	∞_4	$\infty_5^{0/5}$	1	13	13	13	12		10	13	13	13	14		4
14		∞_1	$\infty_2^{0/2}$	1	∞_4	0	1	14	14	14	15		9	14	14	13			7
15		∞_1	0	1	∞_4	0	1						16						
16	•	∞_1	0	$\infty_3^{1/3}$	∞_4	0	1	16	16	9		15	16	16	16	11			1

Table A.19: *a-machine* corresponding to specification D11b.

S	M	State variables						TRANSITIONS											
								11	12	21	22	31	32	41	42	51	52	61	62
1	•	∞_1^1	0	0	∞_4	0	∞_6^1	32	1	10		14		1	1	2		24	1
2	•	∞_1^1	0	0	∞_4	∞_5^0	∞_6^1	7	2	9		3		2	2	2	1	23	2
3		∞_1^1	0	∞_3^0	∞_4	∞_5^0	∞_6^1	6	3	4		3	2	3	3	3	14	16	3
4		∞_1^1	∞_2^0	∞_3^0	∞_4	∞_5^0	∞_6^1	5	4	4	3	4	9	4	4	4	11	17	4
5		1	∞_2^0	∞_3^0	∞_4	∞_5^0	∞_6		4	5	6	5	8	5	5	5	12	18	5
6		1	0	∞_3^0	∞_4	∞_5^0	∞_6^1		3	5		6	7	6	6	6	13	19	6
7		1	0	0	∞_4	∞_5^0	∞_6		2	8		6		7	7	7	32	20	7
8		1	∞_2^0	0	∞_4	∞_5^0	∞_6^1		9	8	7	5		8	8	8	31	21	8
9		∞_1^1	∞_2^0	0	∞_4	∞_5^0	∞_6	8	9	9	2	4		9	9	9	10	22	9
10		∞_1^1	∞_2^0	0	∞_4	0	∞_6	31	10	10	1	11		10	10	9		25	10
11		∞_1^1	∞_2^0	∞_3^0	∞_4	0	∞_6^1	12	11	11	14	11	10	11	11	4		26	11
12		1	∞_2^0	∞_3^0	∞_4	0	∞_6^1		11	12	13	12	31	12	12	5		27	12
13		1	0	∞_3^0	∞_4	0	∞_6^1		14	12		13	32	13	13	6		28	13
14		∞_1^1	0	∞_3^0	∞_4	0	∞_6	13	14	11		14	1	14	14	3		15	14
15		∞_1^1	0	∞_3^0	∞_4	0	1	28	15	26		15	24	15	15	16			14
16		∞_1^1	0	∞_3^0	∞_4	∞_5^0	1	19	16	17		16	23	16	16	16	15		3
17		∞_1^1	∞_2^0	∞_3^0	∞_4	∞_5^0	1	18	17	17	16	17	22	17	17	17	26		4
18		1	∞_2^0	∞_3^0	∞_4	∞_5^0	1		17	18	19	18	21	18	18	18	27		5
19		1	0	∞_3^0	∞_4	∞_5^0	1		16	18		19	20	19	19	19	28		6
20		1	0	0	∞_4	∞_5^0	1		23	21		19		20	20	20	29		7
21		1	∞_2^0	0	∞_4	∞_5^0	1		22	21	20	18		21	21	21	30		8
22		∞_1^1	∞_2^0	0	∞_4	∞_5^0	1	21	22	22	23	17		22	22	22	25		9
23		∞_1^1	0	0	∞_4	∞_5^0	1	20	23	22		16		23	23	23	24		2
24	•	∞_1^1	0	0	∞_4	0	1	29	24	25		15		24	24	23			1
25		∞_1^1	∞_2^0	0	∞_4	0	1	30	25	25	24	26		25	25	22			10
26		∞_1^1	∞_2^0	∞_3^0	∞_4	0	1	27	26	26	15	26	25	26	26	17			11
27		1	∞_2^0	∞_3^0	∞_4	0	1		26	27	28	27	30	27	27	18			12
28		1	0	∞_3^0	∞_4	0	1		15	27		28	29	28	28	19			13
29		1	0	0	∞_4	0	1		24										
30		1	∞_2^0	0	∞_4	0	1		25	30	29	27		30	30	21			31
31	•	1	∞_2^0	0	∞_4	0	∞_6^1		10	31	32	12		31	31	8		30	31
32		1	0	0	∞_4	0	∞_6^1		1	31		13		32	32	7		29	32

Table A.20: *a-machine* corresponding to specification D11c.

A.3 Supervisory Structure

S	M	State variables						TRANSITIONS											
								11	12	21	22	31	32	41	42	51	52	61	62
1	•	0	0	0	0	0	0	2						6		11		21	
2		1	0	0	0	0	0		1	3		4		7		12		22	
3		1	1	0	0	0	0				2	5		8		13		23	
4		1	0	1	0	0	0			5			2	9		14		24	
5		1	1	1	0	0	0				4		3	10		15		25	
6		0	0	0	1	0	0	7							1	16		26	
7		1	0	0	1	0	0		6	8		9			2	17		27	
8	•	1	1	0	1	0	0				7	10			3	18		28	
9		1	0	1	1	0	0			10			7		4	19		29	
10		1	1	1	1	0	0				9		8		5	20		30	
11	•	0	0	0	0	1	0	12						16			1	31	
12		1	0	0	0	1	0		11	13		14		17			2	32	
13		1	1	0	0	1	0				12	15		18			3	33	
14		1	0	1	0	1	0			15			12	19			4	34	
15		1	1	1	0	1	0				14		13	20			5	35	
16		0	0	0	1	1	0	17							11		6	36	
17		1	0	0	1	1	0		16	18		19			12		7	37	
18		1	1	0	1	1	0				17	20			13		8	38	
19		1	0	1	1	1	0			20			17		14		9	39	
20		1	1	1	1	1	0				19		18		15		10	40	
21	•	0	0	0	0	0	1	22						26		31			1
22		1	0	0	0	0	1		21	23		24		27		32			2
23		1	1	0	0	0	1				22	25		28		33			3
24		1	0	1	0	0	1			25			22	29		34			4
25		1	1	1	0	0	1				24		23	30		35			5
26		0	0	0	1	0	1	27							21	36			6
27		1	0	0	1	0	1		26	28		29			22	37			7
28		1	1	0	1	0	1				27	30			23	38			8
29		1	0	1	1	0	1			30			27		24	39			9
30		1	1	1	1	0	1				29		28		25	40			10
31		0	0	0	0	1	1	32						36			21		11
32		1	0	0	0	1	1		31	33		34		37			22		12
33		1	1	0	0	1	1				32	35		38			23		13
34		1	0	1	0	1	1			35			32	39			24		14
35		1	1	1	0	1	1				34		33	40			25		15
36		0	0	0	1	1	1	37							31		26		16
37		1	0	0	1	1	1		36	38		39			32		27		17
38		1	1	0	1	1	1				37	40			33		28		18
39		1	0	1	1	1	1			40			37		34		29		19
40		1	1	1	1	1	1				39		38		35		30		20

Table A.21: *a–machine* corresponding to the supervisory structure of the augmented burner system.

A.4   Local Controllers

S	M	State variables						11	12	21	22	31	32	41	42	51	52	61	62	Cnd. ctrbl.	Unc. trns.
1	•	0	0	0	0	0	0	2						43		12		32			
2		1	0	0	0	0	0		1	3		5		42		13		33			
3		1	1	0	0	0	0				2	4		39		14		34			
4		1	1	1	0	0	0				5		3	40		15		35			
5		1	0	1	0	0	0			4			2	6		16		36			
6		1	0	1	1	0	0									7				•	42
7		1	0	1	1	1	0			8			10		16		41	26			
8		1	1	1	1	1	0				7		9		15		40	25			
9		1	1	0	1	1	0			10	8		8		14		39	24			
10		1	0	0	1	1	0		11	9			7		13		38	23			
11		0	0	0	1	1	0	10							12		37	22			
12	•	0	0	0	0	1	0	13						11			1	21			
13		1	0	0	0	1	0		12	14		16		10			2	20			
14		1	1	0	0	1	0				13	15		9			3	19			
15		1	1	1	0	1	0				16		14	8			4	18			
16		1	0	1	0	1	0			15			13	7			5	17			
17		1	0	1	0	1	1			18			20	26			36		16		
18		1	1	1	0	1	1				17		19	25			35		15		
19		1	1	0	0	1	1				20	18		24			34		14		
20		1	0	0	0	1	1		21	19		17		23			33		13		
21		0	0	0	0	1	1	20						22			32		12		
22		0	0	0	1	1	1	23							21		31		11		
23		1	0	0	1	1	1		22	24			26		20		30		10		
24		1	1	0	1	1	1				23		25		19		29		9		
25		1	1	1	1	1	1				26		24		18		28		8		
26		1	0	1	1	1	1			25			23		17		27		7		
27		1	0	1	1	0	1			28			30		36	26			41		
28		1	1	1	1	0	1				27		29		35	25			40		
29		1	1	0	1	0	1				30	28			34	24			39		
30		1	0	0	1	0	1		31	29		27			33	23			38		
31		0	0	0	1	0	1	30							32	22			37		
32	•	0	0	0	0	0	1	33						31		21			1		
33		1	0	0	0	0	1		32	34		36		30		20			2		
34		1	1	0	0	0	1				33	35		29		19			3		
35		1	1	1	0	0	1				36		34	28		18			4		
36		1	0	1	0	0	1			35			33	27		17			5		
37		0	0	0	1	0	0	38							1	11		31			
38		1	0	0	1	0	0		37	39		41			2	10		30			
39	•	1	1	0	1	0	0				38	40			3	9		29			
40		1	1	1	1	0	0				41		39		4	8		28			
41		1	0	1	1	0	0			40			38		5	7		27			
42		1	0	0	1	0	0									10				•	42
43		0	0	0	1	0	0									11				•	42

Table A.22: *a-machine* corresponding to the local controller obtained from specification D4a.

S	M	State variables						Transitions 11	12	21	22	31	32	41	42	51	52	61	62	Cnd. ctrbl.	Unc. trns.
1	•	0	0	0	0	0	0	2						10		20		40			
2		1	0	0	0	0	0		1	3		5		9		19		39			
3		1	1	0	0	0	0				2	4		8		18		38			
4		1	1	1	0	0	0				5		3	7		17		37			
5		1	0	1	0	0	0			4			2	6		16		36			
6		1	0	1	1	0	0			7			9		5	15		35			
7		1	1	1	1	0	0				6		8		42	14		34			
8	•	1	1	0	1	0	0				9	7			41	13		33			
9		1	0	0	1	0	0		10	8			6		2	12		32			
10		0	0	0	1	0	0	9							1	11		31			
11		0	0	0	1	1	0	12						20			10	30			
12		1	0	0	1	1	0		11	13			15	19			9	29			
13		1	1	0	1	1	0					12	14	18			8	28			
14		1	1	1	1	1	0					15	13	17			7	27			
15		1	0	1	1	1	0				14		12	16			6	26			
16		1	0	1	0	1	0				17			19	15		5	25			
17		1	1	1	0	1	0				16			18	14		4	24			
18		1	1	0	0	1	0				19	17			13		3	23			
19		1	0	0	0	1	0		20	18			16		12		2	22			
20	•	0	0	0	0	1	0	19							11		1	21			
21		0	0	0	0	1	1	22						30		40			20		
22		1	0	0	0	1	1		21	23			25	29		39			19		
23		1	1	0	0	1	1				22	24		28		38			18		
24		1	1	1	0	1	1				25		23	27		37			17		
25		1	0	1	0	1	1			24			22	26		36			16		
26		1	0	1	1	1	1			27			29		25	35			15		
27		1	1	1	1	1	1				26		28		24	34			14		
28		1	1	0	1	1	1				29	27			23	33			13		
29		1	0	0	1	1	1		30	28			26		22	32			12		
30		0	0	0	1	1	1	29							21	31			11		
31		0	0	0	1	0	1	32						40	30				10		
32		1	0	0	1	0	1		31	33			35	39		29			9		
33		1	1	0	1	0	1				32	34		38		28			8		
34		1	1	1	1	0	1				35		33	37		27			7		
35		1	0	1	1	0	1			34			32	36		26			6		
36		1	0	1	0	0	1			37				39	35	25			5		
37		1	1	1	0	0	1				36			38	34	24			4		
38		1	1	0	0	0	1				39	37			33	23			3		
39		1	0	0	0	0	1		40	38			36		32	22			2		
40	•	0	0	0	0	0	1	39							31	21			1		
41		1	1	0	0	0	0											38		•	41
42		1	1	1	0	0	0											37		•	41

Table A.23: *a-machine* corresponding to the local controller obtained from specification D5.

S	M	State variables						TRANSITIONS												Cnd. ctrbl.	Unc. trns.
								11	12	21	22	31	32	41	42	51	52	61	62		
1	•	0	0	0	0	0	0	2							10	20		39			
2		1	0	0	0	0	0		1		3		5		9		19	38			
3		1	1	0	0	0	0					2	4		8		18	40			
4		1	1	1	0	0	0					5		3	7		17	37			
5		1	0	1	0	0	0				4			2	6		16	36			
6		1	0	1	1	0	0				7				9	5	15	34			
7	•	1	1	1	1	0	0					6			8	4	14	35			
8	•	1	1	0	1	0	0					9	7			3	13	33			
9		1	0	0	1	0	0			10	8		6			2	12	32			
10		0	0	0	1	0	0	9								1	11	31			
11		0	0	0	1	1	0	12							20			10	30		
12		1	0	0	1	1	0		11		13		15		19			9	29		
13		1	1	0	1	1	0					12	14		18			8	28		
14		1	1	1	1	1	0					15		13	17			7	27		
15		1	0	1	1	1	0				14			12	16			6	26		
16		1	0	1	0	1	0				17			19	15			5	25		
17		1	1	1	0	1	0					16		18	14			4	24		
18		1	1	0	0	1	0					19	17		13			3	23		
19		1	0	0	0	1	0			20	18		16		12			2	22		
20	•	0	0	0	0	1	0	19							11			1	21		
21		0	0	0	0	1	1	22							30		39		20		
22		1	0	0	0	1	1		21		23		25		29		38		19		
23		1	1	0	0	1	1					22	24		28		40		18		
24		1	1	1	0	1	1					25		23	27		37		17		
25		1	0	1	0	1	1				24			22	26		36		16		
26		1	0	1	1	1	1				27				29	25	34		15		
27		1	1	1	1	1	1					26			28	24	35		14		
28		1	1	0	1	1	1					29	27			23	33		13		
29		1	0	0	1	1	1			30	28		26			22	32		12		
30		0	0	0	1	1	1	29								21	31		11		
31		0	0	0	1	0	1	32									39	30	10		
32		1	0	0	1	0	1		31		33		34				38	29	9		
33		1	1	0	1	0	1					32							6	•	42
34		1	0	1	1	0	1				35			32	36		26		6		
35		1	1	1	1	0	1					34							5	•	42
36		1	0	1	0	0	1				37			38	34		25		5		
37		1	1	1	0	0	1					36								•	41
38		1	0	0	0	0	1		39	40			36		32		22		2		
39	•	0	0	0	0	0	1	38									31	21	1		
40		1	1	0	0	0	1					38								•	41

Table A.24: *a-machine* corresponding to the local controller obtained from specification D10a.

S	M	State variables						11	12	21	22	31	32	41	42	51	52	61	62	Cnd. ctrbl.	Unc. trns.
1	•	0	0	0	0	0	0	2						10		20		39			
2		1	0	0	0	0	0		1	3		5		9		19		38			
3		1	1	0	0	0	0				2	4		8		18		40			
4		1	1	1	0	0	0				5		3	7		17		36			
5		1	0	1	0	0	0			4			2	6		16		37			
6		1	0	1	1	0	0			7			9		5	15		35			
7		1	1	1	1	0	0				6		8		4	14		34			
8	•	1	1	0	1	0	0				9	7			3	13		33			
9		1	0	0	1	0	0		10	8		6			2	12		32			
10		0	0	0	1	0	0	9							1	11		31			
11		0	0	0	1	1	0	12							20		10	30			
12		1	0	0	1	1	0		11	13		15			19		9	29			
13		1	1	0	1	1	0				12	14			18		8	28			
14		1	1	1	1	1	0				15		13		17		7	27			
15		1	0	1	1	1	0			14			12		16		6	26			
16		1	0	1	0	1	0			17			19	15			5	25			
17		1	1	1	0	1	0				16		18	14			4	24			
18		1	1	0	0	1	0				19	17		13			3	23			
19		1	0	0	0	1	0		20	18		16		12			2	22			
20	•	0	0	0	0	1	0	19						11			1	21			
21		0	0	0	0	1	1	22						30		39			20		
22		1	0	0	0	1	1		21	23		25		29		38			19		
23		1	1	0	0	1	1				22	24		28		40			18		
24		1	1	1	0	1	1				25		23	27		36			17		
25		1	0	1	0	1	1			24			22	26		37			16		
26		1	0	1	1	1	1			27			29		25	35			15		
27		1	1	1	1	1	1				26		28		24	34			14		
28		1	1	0	1	1	1				29	27			23	33			13		
29		1	0	0	1	1	1		30	28		26			22	32			12		
30		0	0	0	1	1	1	29							21	31			11		
31		0	0	0	1	0	1	32							39	30			10		
32		1	0	0	1	0	1		31	33		35			38	29			9		
33		1	1	0	1	0	1				32	34			40	28			8		
34		1	1	1	1	0	1				35		33		36	27			7		
35		1	0	1	1	0	1						32							•	42
36		1	1	1	0	0	1				37		40	34		24			4		
37		1	0	1	0	0	1						38							•	41
38		1	0	0	0	0	1		39	40		37		32		22			2		
39	•	0	0	0	0	0	1	38						31		21			1		
40		1	1	0	0	0	1				38	36		33		23			3		

Table A.25: *a-machine* corresponding to the local controller obtained from specification D10b.

S	M	State variables						11	12	21	22	31	32	41	42	51	52	61	62	Cnd. ctrbl.	Unc. trns.
												TRANSITIONS									
1	•	0	0	0	0	0	0	2							10		20		33		
2		1	0	0	0	0	0		1	3		5			9		19		34		
3		1	1	0	0	0	0				2	4			8		18		40		
4		1	1	1	0	0	0					5		3	7		17		39		
5		1	0	1	0	0	0				4			2	6		16		38		
6		1	0	1	1	0	0				7			9	5		15		37		
7		1	1	1	1	0	0					6	8		4		14		36		
8	•	1	1	0	1	0	0					9	7		3		13		35		
9		1	0	0	1	0	0			10	8		6		2		12		32		
10		0	0	0	1	0	0	9							1	11			31		
11		0	0	0	1	1	0	12							20	10			30		
12		1	0	0	1	1	0		11	13			15		19	9			29		
13		1	1	0	1	1	0				12	14			18	8			28		
14		1	1	1	1	1	0					15	13		17	7			27		
15		1	0	1	1	1	0				14			12	16	6			26		
16		1	0	1	0	1	0				17			19	15	5			25		
17		1	1	1	0	1	0					16		18	14	4			24		
18		1	1	0	0	1	0					19	17		13	3			23		
19		1	0	0	0	1	0			20	18		16		12	2			22		
20	•	0	0	0	0	1	0	19							11	1			21		
21		0	0	0	0	1	1	22							30	33			20		
22		1	0	0	0	1	1		21	23			25		29	34			19		
23		1	1	0	0	1	1				22	24			28	40			18		
24		1	1	1	0	1	1					25		23	27	39			17		
25		1	0	1	0	1	1				24			22	26	38			16		
26		1	0	1	1	1	1				27			29		25	37		15		
27		1	1	1	1	1	1					26	28		24		36		14		
28		1	1	0	1	1	1					29	27		23		35		13		
29		1	0	0	1	1	1			30	28		26		22		32		12		
30		0	0	0	1	1	1	29							21	31			11		
31		0	0	0	1	0	1	32								33	30		10		
32		1	0	0	1	0	1		31											•	42
33	•	0	0	0	0	0	1	34						31		21			1		
34		1	0	0	0	0	1		33											•	41
35		1	1	0	1	0	1					32	36		40	28			8		
36		1	1	1	1	0	1					37		35	39	27			7		
37		1	0	1	1	0	1				36			32	38	26			6		
38		1	0	1	0	0	1				39			34	37	25			5		
39		1	1	1	0	0	1					38		40	36	24			4		
40		1	1	0	0	0	1					34	39		35	23			3		

Table A.26: *a-machine* corresponding to the local controller obtained from specification D10c.

S	M	State variables						11	12	21	22	31	32	41	42	51	52	61	62	Cnd. ctrbl.	Unc. trns.
1	•	0	0	0	0	0	0	2							10		19	39			
2		1	0	0	0	0	0		1	3		5			9		18	38			
3		1	1	0	0	0	0				2	4			8		40	37			
4		1	1	1	0	0	0				5		3	7			16	36			
5		1	0	1	0	0	0				4			2	6		17	35			
6		1	0	1	1	0	0				7				9	5	15	34			
7		1	1	1	1	0	0					6			8	4	14	33			
8	•	1	1	0	1	0	0					9	7			3	13	32			
9		1	0	0	1	0	0			10	8		6			2	12	31			
10		0	0	0	1	0	0	9								1	11	30			
11		0	0	0	1	1	0	12							19		10	29			
12		1	0	0	1	1	0		11	13			15		18		9	28			
13		1	1	0	1	1	0					12	14		40		8	27			
14		1	1	1	1	1	0						15	13	16		7	26			
15		1	0	1	1	1	0								12					•	42
16		1	1	1	0	1	0					17		40	14		4	23			
17		1	0	1	0	1	0				16			18	15		5	24			
18		1	0	0	0	1	0			19	40		17		12		2	21			
19	•	0	0	0	0	1	0	18							11		1	20			
20		0	0	0	0	1	1	21							29		39		19		
21		1	0	0	0	1	1		20	22			24		28		38		18		
22		1	1	0	0	1	1					21	23		27		37		40		
23		1	1	1	0	1	1					24		22	26		36		16		
24		1	0	1	0	1	1				23			21	25		35		17		
25		1	0	1	1	1	1				26				28	24	34		15		
26		1	1	1	1	1	1					25	27			23	33		14		
27		1	1	0	1	1	1					28	26			22	32		13		
28		1	0	0	1	1	1			29	27		25			21	31		12		
29		0	0	0	1	1	1	28								20	30		11		
30		0	0	0	1	0	1	31									39	29	10		
31		1	0	0	1	0	1			30	32		34				38	28	9		
32		1	1	0	1	0	1					31	33				37	27	8		
33		1	1	1	1	0	1						34	32			36	26	7		
34		1	0	1	1	0	1				33			31			35	25	6		
35		1	0	1	0	0	1				36			38	34			24	5		
36		1	1	1	0	0	1					35		37	33			23	4		
37		1	1	0	0	0	1					38	36		32			22	3		
38	•	1	0	0	0	0	1			39	37		35		31			21	2		
39	•	0	0	0	0	0	1	38							30		20		1		
40		1	1	0	0	1	0					18	16		13			3	22		

Table A.27: *a-machine* corresponding to the local controller obtained from specification D9a.

S	M	State variables						11	12	21	22	31	32	41	42	51	52	61	62	Cnd. ctrbl.	Unc. trns.
1	●	0	0	0	0	0	0	2						10		13		35			
2		1	0	0	0	0	0		1	3			5	9		14		36			
3		1	1	0	0	0	0				2	4		8		15		37			
4		1	1	1	0	0	0				5		3	7		16		38			
5		1	0	1	0	0	0			4			2	6		17		39			
6		1	0	1	1	0	0			7				9	5	18		32			
7		1	1	1	1	0	0					6		8	4	19		31			
8	●	1	1	0	1	0	0					9	7		3	20		40			
9		1	0	0	1	0	0		10	8		6			2	12		33			
10		0	0	0	1	0	0	9							1	11		34			
11		0	0	0	1	1	0	12						13			10	23			
12		1	0	0	1	1	0		11											●	42
13	●	0	0	0	0	1	0	14						11			1	24			
14		1	0	0	0	1	0		13	15		17		12			2	25			
15		1	1	0	0	1	0				14	16		20			3	26			
16		1	1	1	0	1	0				17			15	19		4	27			
17		1	0	1	0	1	0			16				14	18		5	28			
18		1	0	1	1	1	0			19				12	17		6	29			
19		1	1	1	1	1	0					18		20	16		7	30			
20		1	1	0	1	1	0					12	19		15		8	21			
21		1	1	0	1	1	1					22	30		26		40		20		
22		1	0	0	1	1	1			23	21	29			25		33		12		
23		0	0	0	1	1	1	22							24		34		11		
24		0	0	0	0	1	1	25							23		35		13		
25		1	0	0	0	1	1			24	26	28			22		36		14		
26		1	1	0	0	1	1					25	27		21		37		15		
27		1	1	1	0	1	1					28		26	30		38		16		
28		1	0	1	0	1	1				27			25	29		39		17		
29		1	0	1	1	1	1				30			22		28	32		18		
30		1	1	1	1	1	1					29		21		27	31		19		
31		1	1	1	1	0	1					32		40		38	30		7		
32		1	0	1	1	0	1				31			33		39	29		6		
33		1	0	0	1	0	1			34	40		32			36	22		9		
34		0	0	0	1	0	1	33								35	23		10		
35	●	0	0	0	0	0	1	36								34	24		1		
36		1	0	0	0	0	1			35	37			33		25			2		
37		1	1	0	0	0	1					36	38	40		26			3		
38		1	1	1	0	0	1					39		37	31	27			4		
39		1	0	1	0	0	1					38		36	32	28			5		
40		1	1	0	1	0	1					33	31			37	21		8		

Table A.28: *a-machine* corresponding to the local controller obtained from specification D9b.

References

N. Akamatsu, Y. Sakaki, H–J. Brass, M. Tamaoka, and K. Matsunaga. Batch sequence control using a distributed control system with a coordination station. In *Proc. IFAC DYCORD+ '89. Maastricht, The Netherlands*, pages 323–328, 1989.

C. Arthur. Ambulance computer system was "too complicated". *New Scientist*, page 7, 1992. 14 November.

S. Balemi, G. J. Hoffmann, P. Gyugyi, H. Wong–Toi, and G. F. Franklin. Supervisory control of a rapid thermal multiprocessor. *IEEE Transactions on Automatic Control (joint issue with Automatica)*, 38(7):1040–1059, 1993.

P. I. Barton and C. C. Pantelides. The modelling of combined discrete/continuous processes. *AIChE Journal*, 40:966–979, 1994.

A. Benveniste and K. J. Astrom. Meeting the challenge of computer science in the industrial applications of control. *Automatica*, 29(5):1169–1175, 1993.

A. Benveniste, P. LeGuernic, and C. Jacquemont. The SIGNAL software environment for real–time system specification, design and implementation. *IEEE*, pages 41–49, 1989.

G. M. Birtwistle. *Discrete Event Modelling on SIMULA*. MacMillan, London, 1979.

K. S. Brace, R. L. Rudell, and R. E. Bryant. Efficient implementation of a BDD package. In *Proc. 27th ACM/IEEE Design Automation Conference*, pages 40–45, 1990.

B. A. Brandin and W. M. Wonham. Supervisory control of timed discrete–event systems. *IEEE Transactions on Automatic Control*, 39(2):329–531, February 1994.

R. D. Brandt, V. Garg, R. Kumar, F. Lin, S. Marcus, and W. M. Wonham. Formulas for calculating supremal controllable and normal sublanguages. *Systems and Control Letters*, 15:111–117, 1990.

Y. Brave and M. Heymann. Control of discrete event systems modelled as hierarchical state machines. *IEEE Transactions on Automatic Control*, pages 1803–1819, December 1993.

A. Brooks, R. Cieslak, and P. Varaiya. A method for specifying, implementing and verifying media access protocols. *IEEE Control Systems Magazine*, pages 87–94, June 1990.

J. R. Burch, E. M. Clarke, K. L. McMillan, D. L. Dill, and J. Hwang. Symbolic model checking: 10^{20} and beyond. *Proceedings Symposium on Logics and Computation Sciences*, 1990.

P. E. Caines and S. Wang. COCOLOG: a conditional observer and controller logic for finite machines. In *Proceedings of the 30th Conference on Decision and Control, Brighton, UK*, pages 2845–2850, December 1991.

P. E. Caines, T. Mackling, and Y. J. Wei. Logic control via automatic theorem proving: COCOLOG fragments implemented in Blitzentrum 5.0. In *Proceedings of the 1993 American Control Conference, San Francisco, California*, pages 1209–1213, June 1993.

S. Chung, S. Lafortune, and F. Lin. Limited lookahead policies in supervisory control of discrete event systems. *IEEE Transactions on Automatic Control*, 1992.

R. Cieslak, C. Desclaux, A. Fawas, and P. Varaiya. Supervisory control of discrete–event processes with partial observations. *IEEE Transactions on Automatic Control*, 33(3):249–260, 1988.

E. M. Clarke, E. A. Emerson, and A. P. Sistla. Automatic verification of finite state concurrent systems using temporal logic specifications. *ACM Transactions on Programming Languages and Systems*, 8(2), 1986.

D. Comer. *Digital Logic and State Machine Design*. Holt, Rinehart and Winston, Intl. Edition, 1984.

B. J. Cott and S. Macchietto. An integrated approach to computer–aided operation on batch chemical plants. *Computers and Chemical Engineering*, pages 1263–1271, 1989.

C. A. Crooks. *Synthesis of Operating Procedures for Chemical Processes*. PhD thesis, University of London, 1992.

B. A. Davey and H. A. Priestley. *Introduction to lattices and order*. Cambridge University Press, 1990.

M. Diaz and G. Da Silveira. Specification and validation of protocols by temporal logic and nets. In R. E. A. Mason, editor, *Information Processing 83*, pages 47–52. Elsevier Science, 1983.

M. Diaz. Petri nets based models in the specification and verification of protocols. In W. Brauer and W. Reisig, editors, *Petri Nets: Applications and Relationships to other Models of Concurrency*, Advances in Petri Nets 1986. Part II. Lecture Notes in Computer Sciences no. 225, pages 135–170. Springer Verlag, 1987.

E. W. Dijkstra and C. Scholten. *Predicate Calculus and Program Semantics*. Springer Verlag, 1990.

D. L. Dill. *Trace Theory for Automatic Hierarchical Verification of Speed Independent Circuits*. ACM Distinguished Dissertations. MIT Press, 1989.

T. Donellan. *Lattice Theory*. Pergamon Press, 1968.

S. Eilenberg. *Automata, Languages and Machines, Vol. A*. Academic Press, New York, NY, 1974.

E. A. Emerson and E. M. Clarke. Using branching time temporal logic to synthesize synchronization skeletons. *Science of Computer Programming*, 2:241–266, 1982.

E. A. Emerson and J. Y. Halpern. "Sometimes" and "Not Never" revisited: On branching versus linear time temporal logic. *ACM Journal*, 33(1):151–178, January 1986.

A. Fusaoka, H. Seki, and K. Takahashi. A description and reasoning of plant controllers in temporal logic. In *Proceedings of the 8th International Conference on Artificial Intelligence*, volume 1, pages 405–408, 1983.

A. Galton. Temporal logic and computer science: An overview. In A. Galton, editor, *Temporal Logics and their Applications*. Academic Press, London, 1989.

S. Gaubert. Timed automata and discrete event systems. In *Proceedings of the 1993 European Control Conference, Groningen, NL*, pages 2175–2180, June 1993.

C. Ghezzi and M. Pezze. Cabernet: A customizable environment for the specification and analysis of real time systems. *Submitted for Publication*, 1993.

R. Goldblatt. *Logics of Time and Computation*. Center for The Study of Language and Information, 2nd edition, 1992.

The Guardian. 999 breakdown 'linked to deaths'. *The Guardian Newspaper*, page 4, 1992. 28 October.

R. Hale. Using temporal logic for prototyping: The design of a lift controller. In B. Banieqbal, H. Barringer, and A. Pnueli, editors, *Temporal Logic in Specification. Proc. Col. on Temporal Logic in Specification. Apr. 1987, Altricham, UK.*, Lecture Notes in Computer Sciences, 398, pages 375–408. Academic Press, 1989.

J. Halpern, B. Moskowski, and Z. Manna. A hardware semantics based on temporal intervals. In *Proceedings of ICALP'83. Lecture Notes in Computer Sciences 154*, pages 278–291. Springer Verlag, 1983.

H. M. Hanisch. Coordination control modelling in batch production systems by means of Petri Nets. *Computers and Chemical Engineering*, 16(1):1–10, 1992.

H. M. Hanisch. Analysis of place/transition nets with timed arcs and its application to batch process control. In *Proceedings of the International Conference on applications and theory of Petri Nets, Chicago, USA*, 1993.

D. Harel. STATECHARTS: a visual formalism for complex systems. *Science of Computer Programming*, 8:231–274, 1987.

M. A. Harrison. *An introduction to switching and automata theory*. McGraw-Hill, 1965.

M. Heymann and Feng Lin. On–line control of partially observed discrete event systems. In *Proceedings of the 1993 European Control Conference, Groningen, NL*, pages 2169–2174, June 1993.

M. Heymann. Concurrency and discrete event control. *IEEE Control Systems Magazine*, 10(4):103–112, June 1990.

D. Hiranaka and H. Nishitani. Sequential control issues in the plant-wide control system. In *Preprints of IFAC symposium "ADCHEM'94", Advanced Control of Chemical Processes*, pages 361–366, 1994.

Y. Ho and X. Cao. *Perturbation Analysis of Discrete Event Dynamic Systems*. Kluwer Academic, 1992.

Y. C. Ho. Dynamics of discrete event systems. *Proceedings of IEEE*, 77(1):3–6, 1989.

J. E. Hopcroft and J. D. Ullman. *Introduction to Automata Theory, Languages, and Computation*. Addison–Wesley, Reading, MA, USA, 1977.

International Electrochemical Commission. Preparation of function charts for control systems. Technical Report 848, 1988.

ISA-dS88.01. Batch control systems. Models and terminology. Draft 9. Technical report, ISA, 1994.

C. B. Jones, K. D. Jones, P. A. Linsday, and R. Moore. *MURAL: a formal development support system*. Springer–Verlag, 1991.

N. Klittich and W. Seifert. Functional specification of control systems. In *Proceedings ESPRIT 88*, pages 1361–1377, 1988.

J. F. Knight and K. M. Passino. Decidability for a temporal logic used in discrete event system analysis. *Intl. J. Control*, 52(6):1489–1506, 1990.

R. Kumar, V. Garg, and S. Marcus. On controllability and normality of discrete event dynamical systems. *Systems and Control Letters*, 17:157–168, 1991.

R. Kumar, V. Garg, and S. Marcus. Using predicate transformers for supervisory control. *Proceedings of the 30th Conference on Decision and Control, Brighton, UK*, pages 98–103, December 1991.

Y. Li and W. M. Wonham. Control of vector discrete–event systems. i.– the base model. *IEEE Transactions on Automatic Control*, 38(3):1215–1227, August 1993.

Y. Li and W. M. Wonham. Control of vector discrete–event systems. ii.– Controller synthesis. *IEEE Transactions on Automatic Control*, 39(3):512–531, March 1994.

F. Lin and W. M. Wonham. Decentralized control and coordination of discrete-event systems with partial observations. *IEEE Transactions on Automatic Control*, 35(12):1330–1337, December 1990.

P. A. Lindsay. A survey of mechanical support for formal reasoning. *Software Engineering Journal*, pages 3–27, January 1988.

J. A. Lynch. Industrial exemplars in FOREST research: Requirement specification of an aircraft hydraulics system isolation valve controller. Technical Report NFR/WP4.1/BAe/RP/008, British Aerospace Limited, Warton Aerodrome, Preston, Lancashire PR4 1AX, UK, October 1991.

Z. Manna and A. Pnueli. Verification of concurrent programs: The temporal framework. In R.S. Boyer and J.S. Moore, editors, *The Correctness Problem in Computer Science. International Lecture Series in Computer Science*, pages 215–273. Academic Press, London, 1982.

Z. Manna and A. Pnueli. Verification of concurrent programs: A temporal proof system. Technical Report STAN-CS-83-967, Department of Computer Science, Stanford University, June 1983.

Z. Manna and P. Wolper. Synthesis of communicating processes from temporal logic specifications. *ACM Transactions on Programming Languages*, 6(1):68–93, 1984.

P. Marks. Faults highlight problems of nuclear software. *New Scientist*, page 19, 1992. 29 August.

A. McIver. A question of identity. *Nature*, 368:589–590, 1994. 14 April.

G. Michel. *Programmable Logic Controllers. Architecture and Applications*. John Wiley and Sons, 1990.

I. Moon and S. Macchietto. Formal verification of batch processing control procedures. In *Proceedings of the PSE'94 Kyongju, Korea*, pages 469–475, 1994.

I. Moon, G. Powers, J. R. Burch, and E. M. Clarke. Automatic verification of sequential control systems using Temporal Logic. *AIChE Journal*, 38(1):67–75, January 1992.

J. S. Ostroff. Synthesis of controllers for real–time discrete event systems. *Proceedings of the 28th Conference on Decision and Control, Tampa, Florida*, pages 138–144, December 1989.

J. S. Ostroff. *Temporal Logic for Real-Time Systems*. Research Studies Press/Wiley, 1989.

C. M. Ozveren and A. S. Willsky. Aggregation and multi-level control in discrete event dynamic systems. *Automatica*, 28(3):565–577, 1992.

K. M. Passino and P. J. Antsaklis. Branching time temporal logic for discrete event system analysis. *Intl. J. Control*, 52(6):1489–1506, 1990.

S. S. Pinter and P. Wolper. A temporal logic for reasoning about partially ordered computations. In *Proceedings of the 3rd ACM Symposium on Principles of Distributed Computing. Vancouver, CA*, pages 28–37, 1984.

A. Pnueli. The temporal logic of programs. In *Proceedings of the 8th Symposium on Foundations of Computer Science. Providence, RI*, pages 46–57, 1984.

A. Potton. *An Introduction to Digital Logic*. McMillan, 1973.

H. A. Preisig. The application of Finite Automata Theory to sequential control of chemical processes. *IFAC, DYCORD+ '89, Maastricht, The Netherlands*, pages 99–106, 1989.

A. A. B. Pritsker. *Introduction to simulation and SLAMII*. Halsted, New York, 3rd edition, 1986.

J. Prock. A new technique for fault detection using Petri Nets. *Automatica*, 27(2):239–245, 1991.

P. J. Ramadge and W. M. Wonham. Modular feedback logic for discrete event systems. *SIAM Journal of Control and Optimization*, 25(5):1202–1218, 1987.

P. J. Ramadge and W. M. Wonham. Supervisory control of a class of discrete-event processes. *SIAM Journal of Control and Optimization*, 25(1):206–230, 1987.

P. J. Ramadge and W. M. Wonham. The control of discrete event systems. *Proc. IEEE*, 77(1):81–97, January 1989.

A. P. Ravn, H. Rischel, and K. M. Hansen. Specifying and verifying requirements of real-time systems. *IEEE trans. on software engineering*, 19(1):41–55, 1993.

W. Reisig. Petri Nets in Software Engineering. In W. Brauer and W. Reisig, editors, *Petri Nets: Applications and Relationships to other Models of Concurrency*, Advances in Petri Nets 1986. Part II. Lecture Notes in Computer Sciences no. 225, pages 63–96. Springer Verlag, 1987.

J. R. Rivas and D. F. Rudd. Synthesis of failure–safe operations. *AIChE J*, 20:320–325, 1974.

R. Y. Rubenstein and A. Shapiro. *Discrete event systems: sensitivity analysis and stochastic optimization by the score function method*. Wiley, Chichester, UK, 1992.

K. Rudie and J. C. Willems. The computational complexity of decentralized discrete–event control problems. In *Proceedings of the 1993 European Control Conference, Groningen, NL*, pages 2185–2190, June 1993.

P. Sawyer. *Computer Controlled Batch Processing*. IChemE, Rugby, UK, 1993.

T. J. Schriber. *Simulation using GPSS*. Wiley, New York, 1974.

H. Schuler, F. Allgower, and E. D. Gilles. Chemical Process Control: Present status and future needs. The view from the European industry. In Y. Arkun and W. H. Ray, editors, *CPC IV*, pages 29–52. Elsevier, 1991.

J. G. Thistle and W. M. Wonham. Control problems in a temporal logic framework. *Intl. J. Control*, 44(4):943–976, 1986.

J. P. Tremblay and Z. Manohar. *Discrete Mathematical Structures with Applications to Computer Science*. McGraw-Hill, 1975.

J. Tsitsiklis. On the control of discrete–event dynamical systems. *Mathematics of Control, Signals and Systems*, 2:95–107, 1989.

P. Varaiya and A. B. Kurshanzki. *Discrete Event Systems: Models and Applications*. Springer Verlag, 1988.

F. Wang and G. N. Saridis. A coordination theory for intelligent machines. *Automatica*, 2(5):833–844, 1990.

H. Wang. Non–interleaving approach to the modeling and specification issues of concurrent DES. In *Proceedings of the 1993 American Control Conference, San Francisco, California*, pages 1209–1213, June 1993.

M. J. Wilkins. Simplify batch automation projects. *Chemical Engineering Progress*, pages 61–66, April, 1992.

H. P. Williams. *Model Building in Mathematical Programming*. Wiley, 3rd edition, 1990.

R. G. Willson and B. H. Krogh. Petri nets for the specification and analysis of discrete controllers. *IEEE Transactions on Software Engineering*, 16(1):39–50, 1990.

P. Wolper. Temporal logic can be more expressive. *Information and Control*, 56(1–2):72–99, 1983.

P. Wolper. On the relation of programs and computations to models of temporal logic. In B. Banieqbal, H. Barringer, and A. Pnueli, editors, *Temporal Logic in Specification. Proc. Col. on Temporal Logic in Specification. Altricham, UK, Apr. 1987*, Lecture Notes in Computer Sciences, 398, pages 75–123. Academic Press, 1989.

H. Wong–Toi and G. Hoffmann. The control of dense real-time discrete event systems. In *Proceedings of the 30th Conference on Decision and Control, Brighton, UK*, pages 1527–1528, December 1991.

W. M. Wonham and P. J. Ramadge. On the supremal controllable sublanguage of a given language. *SIAM Journal of Control and Optimization*, 25(3):637–659, 1987.

W. M. Wonham and P. J. Ramadge. Modular supervisory control of discrete event systems. *Mathematics of Control, Signals and Systems*, 1(1):13–30, 1988.

W. M. Wonham. A control theory for discrete–event systems. In M.J. Denham and A.J. Laub, editors, *Advanced Computing Concepts and Techniques in Control Engineering*, pages 129–169. Springer Verlag, 1988.

W. M. Wonham. On the control of discrete–event systems. In H. Nijmeijer and J. M. Schumacher, editors, *Three decades of Mathematical System Theory. Lecture Notes in Control and Information, vol. 135*, pages 542–562. Springer Verlag, 1989.

E. C. Yamalidou and J. C. Kantor. Modeling and optimal control of discrete–event chemical processes using Petri Nets. *Computers and Chemical Engineering*, 15(7):503–519, 1991.

E. C. Yamalidou, E. P. Patsidou, and J. C. Kantor. Modelling discrete–event dynamical systems for chemical process control – A survey of several new techniques. *Computers and Chemical Engineering*, 14(3):281–299, 1990.

E. C. Yamalidou, E. D. Adamides, and D. Bonvin. Optimal failure recovery in batch processing using Petri Net models. In *Proceedings of the 1992 American Control Conference*, pages 1906–1910, 1992.

H. Zhong and W. M. Wonham. On the consistency of hierarchical supervision in discrete–event systems. *IEEE Transactions on Automatic Control*, 35(10):1125–1134, October 1990.

Glossary

a–machine	A labelled finite state machine with only one initial state in which all the states are labelled by the status of the system being modelled.
alphabet	A set of symbols.
Boolean lattice	A complemented, distributive lattice.
bounded lattice	A finite lattice.
complemented lattice	A lattice in which every element in the lattice has at least one complement.
distributive lattice	A lattice satisfying the distributive law.
homomorphism	A mapping h between any two algebraic systems X, Y of the same type, such that $h : X \to Y$, in which their operations are preserved.
language	Set of all finite strings that can occur within a string set.
lattice	A partially ordered set $< S, \leq >$ in which every pair of elements $a, b \in S$ has a greatest lower bound (glb) and a least upper bound (lub). See main text for an alternative definition as an algebraic system.
marked language	All strings formed from an initial state $q_0 \in Q$ where the final state is a member of the set of marked states Q_m in a given finite state machine.
Mealy machine	A finite machine $m = \{Q, \Sigma, O, \delta, \phi, q_0\}$, where Q, set of states; Σ, set of transitions; O, set of outputs;

δ, partial function $\delta : Q \times \Sigma \to Q$; ϕ, partial function $\phi : Q \times \Sigma \to O$; q_0, initial state.

partial order

A binary relation R in a set S iff R is reflexive, antisymmetric, and transitive.

power set

The power set of set S, $P(S)$, consists of all the subsets of S ordered by inclusion, that is: for $A, B \in P(S)$, $A \leq B$ is defined iff $A \subseteq B$.

predicate transformer

A function from boolean structures to boolean structures.

prefix

Any number of leading symbols in a string.

prefix closure language

The language which possesses all the prefixes of the strings of a given language L.

quasi–lattice

A lattice representing a process with self-loops attached to lattice nodes.

regular language

A language that can be represented by a finite state machine.

string

A sequence of symbols.

symbol

Minimum abstract entity in Language Theory which does not need to be defined.

word

A string.

Index

— T —

— W —

Lecture Notes in Control and Information Sciences

Edited by M. Thoma

Vol. 197: Henry, J.; Yvon, J.P. (Eds)
System Modelling and Optimization
975 pp approx. 1994 [3-540-19893-8]

Vol. 198: Winter, H.; Nüßer, H.-G. (Eds)
Advanced Technologies for Air Traffic Flow
Management
225 pp approx. 1994 [3-540-19895-4]

Vol. 199: Cohen, G.; Quadrat, J.-P. (Eds)
11th International Conference on
Analysis and Optimization of Systems –
Discrete Event Systems: Sophia-Antipolis,
June 15–16–17, 1994
648 pp. 1994 [3-540-19896-2]

Vol. 200: Yoshikawa, T.; Miyazaki, F. (Eds)
Experimental Robotics III: The 3rd
International Symposium, Kyoto, Japan,
October 28-30, 1993
624 pp. 1994 [3-540-19905-5]

Vol. 201: Kogan, J.
Robust Stability and Convexity
192 pp. 1994 [3-540-19919-5]

Vol. 202: Francis, B.A.; Tannenbaum, A.R.
(Eds)
Feedback Control, Nonlinear Systems,
and Complexity
288 pp. 1995 [3-540-19943-8]

Vol. 203: Popkov, Y.S.
Macrosystems Theory and its Applications:
Equilibrium Models
344 pp. 1995 [3-540-19955-1]

Vol. 204: Takahashi, S.; Takahara, Y.
Logical Approach to Systems Theory
192 pp. 1995 [3-540-19956-X]

Vol. 205: Kotta, U.
Inversion Method in the Discrete-time
Nonlinear Control Systems Synthesis
Problems
168 pp. 1995 [3-540-19966-7]

Vol. 206: Aganovic, Z.;.Gajic, Z.
Linear Optimal Control of Bilinear Systems
with Applications to Singular Perturbations
and Weak Coupling
133 pp. 1995 [3-540-19976-4]

Vol. 207: Gabasov, R.; Kirillova, F.M.;
Prischepova, S.V.
Optimal Feedback Control
224 pp. 1995 [3-540-19991-8]

Vol. 208: Khalil, H.K.; Chow, J.H.;
Ioannou, P.A. (Eds)
Proceedings of Workshop on Advances in
Control and its Applications
300 pp. 1995 [3-540-19993-4]

Vol. 209: Foias, C.; Özbay, H.;
Tannenbaum, A.
Robust Control of Infinite Dimensional
Systems: Frequency Domain Methods
230 pp. 1995 [3-540-19994-2]

Vol. 210: De Wilde, P.
Neural Network Models: An Analysis
164 pp. 1996 [3-540-19995-0]

Vol. 211: Gawronski, W.
Balanced Control of Flexible Structures
280 pp. 1996 [3-540-76017-2]